DEDICATION

I would like to dedicate this dissertation to my late grandparents Mrs. Kamla Gupta and Mr. Basudeo Gupta.

ACKNOWLEDGEMENTS

First and foremost I would like to express my gratitude for my mentor Dr. R.K. Rao and my graduate committee members Dr. Don Thomason, Dr. Anjapravanda P. Naren, Dr. Christopher M. Waters and Dr. Edwards Park for their continual guidance and support during my research. I would also like to thank all my lab members for their help and support.

Thanks to the department of physiology office staff for their help and cooperation. Dr. Pat Ryan and Dr. Don Thomason, who in their administrative capacities at the College of Graduate Health Sciences have been very generous with their encouragement and support for me whenever required. My sincere thanks go to Ms. Janie Vanprooijen of Integrated program in biomedical sciences. I would also like to thank Mr. Larry Tague for his continued help and guidance. I would like to thank Ms. Shirley Hancock and her group for helping in format review of this dissertation.

Finally I would like to thank my friends and family for their unflinching support throughout.

ABSTRACT

Occludin is hyperphosphorylated on Ser and Thr residues in intact epithelial tight junctions. The dynamics of epithelial tight junctions appear to involve reversible phosphorylation of occludin on Ser and Thr residues. In the present study we determined the role of PKCζ in occludin phosphorylation and the dynamics of tight junctions. Inhibition of PKCζ by specific PKCζ pseudo substrate rapidly reduced TER, increased inulin permeability and induced redistribution of occludin and ZO-1 in Caco-2 and MDCK cell monolayers without inducing cytotoxicity. Reduced expression of PKCζ also resulted in compromised tight junction integrity. Both PKCζ pseudo substrate and reduced expression of PKCζ delayed the calcium-induced assembly of tight junctions. PKCζ pseudo substrate rapidly reduced occludin phosphorylation on Ser and Thr, and prevented Ser/Thr phosphorylation of occludin during the assembly of tight junctions. Pairwise binding studies showed that PKCζ directly bound to and phosphorylated the C-terminal tail of occludin on Ser and Thr residues. Site directed point mutation demonstrated that PKCζ predominantly phosphorylated T438, but also phosphorylated T403 and T404. This study demonstrates that PKCζ phosphorylated occludin on Ser/Thr residues and regulated the assembly of tight junctions.

TABLE OF CONTENTS

LIST OF FIGURES

CHAPTER 1: TIGHT JUNCTIONS

1.1 Intercellular Junctions

A defining characteristic of all organisms is their ability to respond to stimuli, growth and development, propagation of their own kind (reproduction) and maintenance of a steady-state (homeostasis). The multicellular organisms (eukaryotes) have a more complex cellular structure as compared to the relatively simple unicellular organisms (prokaryotes). Over time the multicellular organisms have evolved to have differentiated cells which can form organs and tissues and have specialized functions. One of the most important functions is to prevent the internal environment from external insults and injury. This is achieved by maintaining homeostasis by establishing an internal milieu distinct from the external environment.[1] Epithelium and endothelium form the barrier function in different organs. This barrier function is important to maintain different physiological compartments in a multi-organ system and also maintain the interior milieu/homeostasis.[2-4] Epithelial barrier function prevents the internal body compartments from external environment whether it be infective agents, chemical agents or physical injury. The skin, blood brain barrier, renal epithelium and intestinal epithelium are examples of such barriers across different organ systems. Breakdown of barrier function leads to changes in the homeostatic state of these organ systems and could contribute to disease pathogenesis.[5]

Epithelial and endothelial cells form selective barriers between different body compartments and tissues as well as extrinsic environmental factors. In order to perform the barrier function, these cells form complexes and adhere to each other by forming junctional complexes with each other.[6] These intercellular junctions in the vertebrate cells consist of the tight junctions (zonula occludens), desmosomes (macula adherens), adherens junctions (zonula adherens) and gap junctions (nexus).[7] The first three are together referred to, as the epithelial junctional complex[6] (**Figure 1.1**).

Before the advent of electron microscopy, not much was known about this complex. However, based on physiological evidence of a 'barrier function', an intercellular seal was suspected. The epithelial cells were thought to have an absolute barrier which was termed as the terminal bar.[2,8] It was thought to be formed from intercellular secretions and their main function was supposed to be cell-cell adhesion rather than regulation of epithelial permeability.[2] Farquhar and Palade first described the tripartite junctional complex that is found between two adjacent cells. They observed that even though the exact arrangement of the complex differs from one organ to the other it was ubiquitous in its distribution throughout the body.[9] The universal distribution of this junctional complex across species and organs further underlines their significance for existence of multicellular life forms.

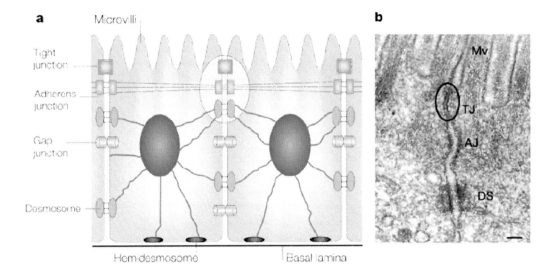

Figure 1.1: Epithelial junction complex.
(a) Schematic drawing of intestinal epithelial cells. (b) Electron micrograph of the junctional complex in mouse intestinal epithelial cells. Reprinted with permission. Tsukita S, Furuse M, Itoh M. Multifunctional strands in tight junctions. Nat Rev Mol Cell Biol 2001;2(4):285-293.[6]

1.1.1 Tight Junctions

Tight junctions are the most apical members of the epithelial junction complex. They are located at the luminal end of the intercellular space. They form a barrier which prevents passage of pathogens, solutes across the mucosal layer into the bloodstream. However, they are not absolute barriers but are semipermeable[10] and regulated[11] diffusion barriers across epithelium. They also play a role in maintaining the polarity of epithelial cells by forming a fence around the cells and separate the apical and basolateral compartment.[12] On transmission electron microscopy they can be seen as dots between adjoining cells.[9] On freeze fracture electron microscopy they are seen as encircling bands around the apicolateral margins of cells.[13] Also they are seen as continuous anastomosing strands between different cells[14] (**Figure 1.2**). The tight junctions are composed of various proteins that may be divided into three groups: the integral proteins, plaque proteins and regulatory proteins.[14]

1.1.2 Adherens Junctions

Adherens junctions are present between two adjoining cell surfaces, situated basally to the tight junctions. They form a continuous adhesion belt. They either encircle the cells in the form of Zonula adherens or occur as points of attachment to the extracellular matrix called adhesion plaques. Adherens junctions are responsible for

2

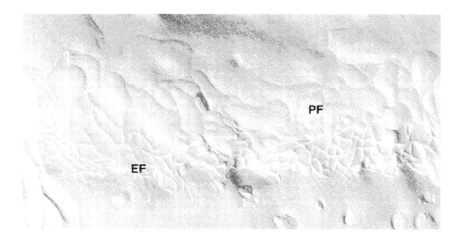

Figure 1.2: Freeze fracture image of tight junctions between bronchial epithelial cells in the rat lung.
PF-protoplasmic face, EF-ectoplasmic face. Reprinted with permission. Mitic LL, Anderson JM. Molecular architecture of tight junctions. Ann Rev Phys 1998; 60:121-142.[147]

keeping adjacent cells together through attachments to actin filaments. They are also known to play a role in contact inhibition.[15-17]

1.1.3 Desmosomes

Desmosomes are spot like adhesions and are present at the junctional surfaces of the adjacent cells. They are randomly distributed along the surface and protect the cells from shearing forces. Desmosomes are responsible for keeping adjacent cells together and provide attachment sites for intermediate filaments.[18,19]

1.1.4 Gap Junctions

Gap junctions are present between adjoining cells and connect the cytoplasm of two cells. They allow for movement of intracellular signaling molecules from one cell to another and also provide cell-cell adhesion.[20,21]

In this study, our focus is on studying the tight junctions. So, the following sections contain a detailed description of the tight junctions.

1.2 Morphology of Tight Junctions

Among the vertebrates, there are various organs which require a separation of environment between the apical and basolateral sides of the epithelial cell surfaces. A few examples are the gastrointestinal tract, urinary tract, blood-brain barrier in the central nervous system and blood-testis barrier in the male reproductive system etc. In the present study, we have tried to look at the tight junctions in the intestinal epithelium and the renal epithelium.

In the intestinal epithelium, mucosa performs an important function of regulating the selective passage of nutrients from the luminal side into the blood stream, and preventing the passage of toxins, pathogens, allergens and other unwanted substances across the epithelium.[22] In the renal tubular epithelium, again the significant function of regulation of passage of solutes and water between the tubules and the renal interstitium is performed by the epithelial cells and the junctions situated in between the adjacent cells. Another interesting fact is that even within the renal tubules, the characteristics of the epithelial cells to allow or restrict the passage of substances vary between the proximal and distal convoluted tubules.[23-25]

These facts including numerous other characteristics of various epithelia, the selective passage of substances across these epithelial layers, and their regulation has evinced considerable interest among the scientific community to study the structure and function of tight junctions. The earlier investigators had proposed a 'lipid model' or a 'protein model' to describe the structure of tight junctions.[26] But with the advances in cell biology, it is now known that tight junctions consist of at least 50 different structural and signaling proteins. Thus, in addition to the structural proteins, the signaling molecules have become a whole new area of interest for scientists involved in tight junction research.

Several of the details of tight junction structure were obscure from us in the era of light microscopy. But electron microscopy has made it possible for us to study their structure in the greatest possible detail. As explained before, both transmission electron microscopy and freeze-fracture electron microscopy have been used to study tight junctions.

The new research in the field has shattered the old myth that tight junctions are rigid structures made up of some proteins or molecules and connecting the adjacent cells by calcium links. We now know that tight junctions are dynamic structures and their tightness of structure and function depends upon the organ system they are present in. This variation or dynamics is understood to be dependent upon number of sealing strands that constitute the tight junction between any two adjacent cells. Moreover, presence of numerous signaling molecules in the vicinity of these sealing strands, and their proven role in the regulation of tight junction structure and function has only given credence to the theory of dynamic tight junctions.

4

1.3 Tight Junction Components

The TJ complex consists of three elements (**Figure 1.3**):

a) *Integral/transmembrane proteins-* Occludin, Claudins and junctional adhesion molecule (JAM). The integral proteins consist of the transmembrane spanning domains which form the basic architecture of the TJs. Also they have cytoplasmic domain(s) which anchor them to the actin cytoskeleton via the scaffolding proteins.[6,23,27-29]

b) *Scaffolding proteins-* Zonula Occludens-1(ZO-1), ZO-2, and ZO-3. These provide an anchor for the transmembrane proteins and connect them to the actin cytoskeleton.[6,23,27-29]

c) *Regulatory proteins-* G-proteins, Rho GTPases, Cingulin, PTEN, protein kinases and phosphatases. These proteins regulate the signaling to and from TJs and maintain TJ permeability.[2,23,27-36]

Newer findings suggest that there maybe a fourth class of proteins which mediate membrane vesicle targeting. All of these proteins act together to couple TJ proteins with the actin cytoskeleton and thus function to control cell permeability, polarization etc. The following sections include a detailed description of various proteins that constitute the tight junction complex.

1.3.1 Scaffolding Proteins

These are the peripheral proteins which function to organize the integral proteins and connect them to cytoplasmic proteins and the actin cytoskeleton. They are also involved in signal transduction at tight junctions. These include ZO-1, ZO-2, ZO-3, symplekin, ZA-1 TJ, AF-6 and cingulin.

1.3.1.1 Zona occludens

Zona occludens proteins belong to the Membrane Associated Guanylate Kinase (MAGUK) family of proteins. Structurally, they are known to contain one SH3 (Src-homology 3) domain, at least one PDZ domain and a Guanyl kinase (GUK) homologous region.[34,37] ZO-1 was the first tight junction protein identified. ZO-1 is a 220-kDa protein that forms a stable complex along with ZO-2 and ZO-3. It binds directly to a 150 amino acid domain at the C-terminal tail of occludin at the cytoplasmic end of occludin.[38] It also binds to the actin binding protein spectrin. ZO-1, ZO-2 and ZO-3 are expressed and colocalize with occludin at the tight junction. They are known to be associated with the plasma membrane. They are also involved in transcription and may influence cell differentiation.[39-43]

5

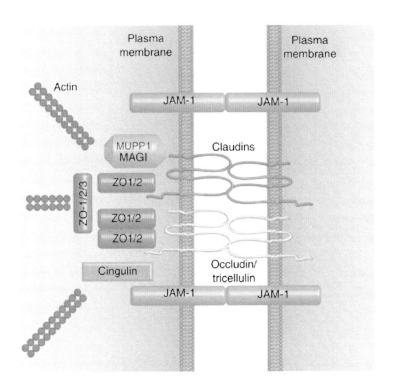

Figure 1.3: Components of tight junction.
Niessen CM. Tight Junctions/Adherens Junctions: Basic Structure and Function.
Retrieved from http://www.nature.com/jid/journal/v127/n11/full/5700865a.html.
Accessed on 06 May 2011.[148]

1.3.1.2 Cingulin

Cingulin is a phosphoprotein with a molecular weight of 140-160 kDa. It forms coiled-coil parallel dimers on the cytoplasmic surface of tight junctions.[44] Further intermolecular aggregations between these coiled-coil dimers lead to their aggregation and interaction with ZO-1 and ZO-2 in the cells. Cingulin has a head domain and a tail domain, both of which are essential in its junctional localization. The functional importance of cingulin in tight junctions lies in its role in inter-connection between the plaque proteins and actin cytoskeleton.[45,46]

1.3.1.3 Symplekin

This 126 kDa protein is also present in the junctional plaque. Symplekin is present in the tight junctions in sertoli cells of the testes. It is, however, not known to be present in the endothelial cells. Besides the tight junctions, Symplekin is also present in the nuclei of the epithelial cells. It is believed to be involved in control of nuclear events relating to tight junctions, like reporting the functional status of tight junction contacts.[47]

1.3.1.4 7H6 Protein

7H6 is a 155 kDa phosphoprotein, and is known to be found in tight junctions of endothelial cells and hepatocytes. It responds to the depletion of ATP in the cells and gets reversibly dissociated from the junctional complex. Thus its role in regulation of tight junctions is dependent upon the functional status of the cells. 7H6 regulates the barrier function in both epithelial and endothelial cells. When the cells are undergoing changes from normal to dysplastic to carcinomatous, the level of expression of 7H6 in the cells decreases gradually and progressively.[48]

1.3.2 Perijunctional Actin

Actin cytoskeleton is connected to the tight junction proteins though the plaque proteins. It has been suggested that the integral proteins are linked to a perijunctional F-actin ring.[40,49,50] Changes in actin organization have been correlated with permeability changes at the epithelial barrier. A suggested model for involvement of perijunctional actin in regulating epithelial permeability is by myosin generated cytoskeletal traction on the apical junction complex. This has been shown to be mediated by myosin light chain kinase (MLCK) which can activate myosin ATPase and contraction of the actin cytoskeleton.[51,52] Studies have shown that increased MLCK activity leads to an attenuation of the epithelial barrier function. Actin binding to TJ is suggested through occludin-ZO-1-ZO-2-F-actin and occludin-ZO-1-ZO-3-F-actin complexes. The actin cytoskeleton not only provides support to the TJ complex but is also involved in various signaling mechanisms that regulate the tight junction.

1.3.3 Integral/Transmembrane Proteins of the Tight Junction

Transmembrane proteins of the tight junction are occludin, claudin and junctional adhesion molecule (JAM) (**Figure 1.4**). Since my thesis topic deals with occludin phosphorylation that will be described after the other proteins in more detail.

1.3.3.1 Junctional adhesion molecules

JAMs belongs to the IgG superfamily of proteins of which four members JAM-1, JAM-2, JAM-3 and JAM-4 have been identified. JAM-1 is a glycosylated protein weighing 43-kDa. It consists of two extracellular domains, one intracellular carboxyl domain and one transmembrane domain. JAM-1 is expressed in both the epithelial and endothelial tight junctions and is supposed to help the monocyte migration. JAM-2 is found in endothelial cells and JAM-3 on T cells. JAM-4 is expressed in renal glomeruli and intestinal epithelial cells. Evidence suggests that JAM-1 is associated to claudins indirectly through ZO-1. JAM-1 also recruits PAR-3/aPKC/Par-6 complex to the tight junction complex.[31,53]

1.3.3.2 Claudins

Claudins were characterized by the Tsukita group.[54] Over the years, the number of members of claudin superfamily has kept on increasing and now it is suggested to have 24 members. Claudins are tetraspan molecules with two extracellular loops, one intracellular loop and cytosolic C- and N- terminus. They also have a PDZ bindng motif at the C-terminus. They interact with matrix proteins like ZO-1 through this motif. Claudins form heteromeric complexes within the same cell and also at the intercellular junction.[55-58] They show a great deal of variability in their sequence, and based on their sequence analyses, Claudins are classified into two groups. They are classic claudins (1-10, 14, 15, 17 and 19) and non-classic claudins (11-13, 16, 18 and 20-24). Even though claudins form an integral part of the tight junction, their expression can vary in different tissues and organs. Claudin-1 is expressed ubiquitously in the cells. Claudin-5 has predominant expression in tight junctions of endothelial cells, claudin 11 in sertoli cells and claudin-16 in the thick ascending loop of henle in the kidney.[4,59] Not all claudins have a 'sealing' function. Claudin-2 and claudin-10A are examples of pore forming claudins. Examples of sealing claudins are claudin-1, 4, 5, 8, 11, 14 and 19. Moreover, each claudin has its unique expression pattern between the crypts and the villi of the mucosa.[4,60] Claudins are believed to be playing a significant role in the maintenance of the structural integrity of tight junctions in the absence of occludin.[61]

1.3.3.3 Occludin

Occludin was also identified and characterized by the Tsukita group.[62] Human occludin has been characterized as a 65 kDa protein with two extracellular loops and two

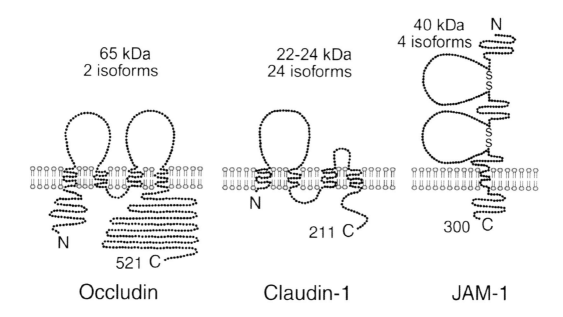

Figure 1.4: The proteins forming the tight junction complex.
JAM, Junctional Adhesion Molecule. Reprinted with permission. Schneeberger EE, Lynch RD. The tight junction: a multifunctional complex. Am J Physiol 2004;286:1213-1228.[149]

cytosolic amino and carboxy termini. The extracellular loops have 46 and 48 amino acids and are separated by a 10 amino acid cytosolic loop. The first extracellular loop has a high number of conserved glycine and tyrosine residues. The N terminus and the C terminus have 65 and 255 amino acids respectively. Both the N and C terminal domains have a number of serine and threonine residues. The C terminal tail of occudin interacts with the plaque proteins ZO-1, ZO-2 and ZO-3. There is a highly conserved 150 amino acid domain at the C terminus that interacts with these proteins.[40,63] Occludin expression can be correlated with the barrier function in various tissues and organs.[64] In studies with occludin homozygous null mice it has been shown that occludin is not essential to form the tight junctions and the mice were viable. However, these mice exhibited abnormalities in the salivary gland, testis, bone and stomach lining.[65] Growth retardation was observed in occludin deficient mice, the males showed sterility and the females could not suckle their offspring. It has been suggested that other tight junction proteins compensate for the lack of occludin expression and yet it is indispensable for a number of functions.[61,66]

Occludin interacts with ZO-1, ZO-2 and ZO-3 at the C terminus, which are further coupled to the actin cytoskeleton. Occludin also interacts with the connexin-32, which is a gap junction protein.[67] Occludin also forms a coiled-coil domain by homodimerizing with itself.[67] It has been suggested that this domain interacts with ZO-1 and also a number of regulatory proteins. It has been shown that the extracellular loops of occludin are involved in the formation of cell-cell contacts and thus creating the paracellular barrier.[68] Occludin has also been shown to be involved in regulation of cell polarity. Occludin overexpression has been shown to increase TER and also decrease paracellular flux in some cell lines, which points to a role of occludin in the formation of paracellular channels.[69,70]

1.3.3.4 Tricellulin

This 70 kDa novel, integral membrane protein is unique in the fact that it is present at tricellular tight junctions where three cells are joined together. Tricellulin is composed of four transmembrane domains and is a component of bicellular and tricellular tight junctions.[71]

1.4 Function of Tight Junctions

Other than forming a barrier[20,72] and regulating the transport of transport of fluids and solutes across the epithelial monolayer,[73,74] tight junctions also have a 'fence function' in biological membranes where they determine cell polarity as apical or basolateral.[75,76] Tight junctions are now known to be involved in regulation of transcription,[77] cell proliferation[78] and differentiation.[78]

Cellular transport consists of transcellular and paracellular pathways[2] (**Figure 1.5**). Transcellular transport is mediated by ion channels and pumps distributed

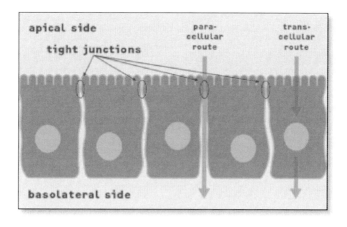

Figure 1.5: Cellular transport.
Schematic drawing of cellular transport showing trans and para cellular routes. Barrier Forming Tissue. Retrieved from http://www.nanoanalytics.de/en/hardwareproducts/cellzscope/howitworks/chapter01/index.php. Accessed on 06 May 2011.[150]

throughout the membrane. It is dependent on availability of energy rich molecules e.g. Adenosine tri-phosphate (active transport). This further gives rise to electric and osmotic gradients across the monolayers. The paracellular pathway is responsible for the passive transport of ions and solutes through the intercellular junctions. The epithelial junction complex is responsible for regulating the paracellular transport.[79] Usually both transcellular and paracellular pathways are present across biological membranes. Transcellular transport occurs as a result of a tightly regulated movement of water and solutes across the epithelial cells and is an active process dependent on the presence of specific channels and transporters. Transcellular transport is therefore very specific and tightly regulated by the physiological state of the cell. The paracellular transport happens as a result of the osmotic gradient generated by the transcellular pathway. The two basic characteristics shown by the paracellular transport are permeability, governed by the magnitude of the barrier function and selectivity, which determines its ability to discriminate solutes or solvents on the basis of ionic charge or molecular size. These two characteristics are determined by tight junctions, which are points of circumferential contacts between adjacent epithelial cells. The tight junction dynamics govern the physiological characteristics of the paracellular pathway.[79]

1.5 Disruption of Barrier Function

Tight junctions are the major determinants of barrier function and loss of tight junction function leads to disruption of barrier function.[5,80] This disruption leads to disease pathogenesis due to loss of homeostasis and exposure of internal body compartments to external environment. Loss of barrier function leads to pathogenesis of various diseases like diarrhea, inflammatory bowel disease (Crohn's disease and

Ulcerative colitis), primary biliary cirrhosis, malabsorption, primary sclerosing cholangitis, celiac disease, endotoxemia and multiple sclerosis etc.[81,84] Loss of the 'fence function' can lead to tumorigenesis and metastasis of cancer cells.[5] Due to the critical role of tight junctions in cell growth and differentiation, and thus regulation of cell proliferation and migration, disruption of tight junctions can cause epithelial-mesenchymal transformation and development of cancers. This latter effect can be produced by both down-regulation and over-expression of the tight junction proteins.[5,36,80-82]

1.6 Tight Junction Regulation

Earlier it was thought that TJ permeability was static but now we know that TJ permeability is affected by extracellular stimuli. This 'plasticity' of tight junctions gives them a more dynamic function where they can be regulated by different signaling mechanisms.[14,74,85] A number of regulatory proteins have been implicated in TJ regulation e.g. G-proteins, Rho GTPases, Cingulin, PTEN, PKCs and protein phosphatases.[2,51] Protein kinases and phosphatases are known to regulate a number of cell functions. The protein kinases act by phosphorylating the substrates and phosphatases by dephophosphorylating the substrates. Thus a balance between phosphorylated and dephosphorylated states of various proteins is maintained and can be shifted either way depending on the state of cellular homeostasis.[86] Protein kinases and phosphatases are also implicated in regulation of TJ permeability. Thus the TJs form a dynamic barrier with inputs from the actin cytoskeleton and regulatory proteins. Tight junction assembly is initiated by the E-cadherin mediated cell-cell adhesion.[4] A number of proteins have been suggested to be involved in the signaling from the E-cadherin mediated cell-cell adhesion to the assembly site for the tight junction e.g. protein kinase A(PKA), monomeric and heterotrimeric G proteins and protein kinase Cs(PKCs).[87] Signals also originate at the tight junctions and are involved in regulation of gene expression, cell proliferation and differentiation.[30,77,88,89] Various regulatory proteins involved in TJ regulation are listed with a brief description below.

1.6.1 Protein Kinase A

PKA has been shown to inhibit denovo assembly of tight junctions in a calcium switch model.[88] But it has also been shown that this effect can be modified in response to other signaling pathways. However the PKA is generally considered to inhibit TJ assembly.[90]

1.6.2 Heterotrimeric G Proteins

G-proteins are present near the inter-cellular junctional complex, in association with tight junction proteins.[91] Tight junction assembly is seen in experiments where inhibitory G-proteins are blocked. However, as with PKA this has been contradicted and overexpression of G-proteins has been shown to stimulate tight junction assembly.[88,91,92]

1.6.3 Cyclic-AMP and Calcium

These second messenger molecules play a significant role in influencing the tight junction structure and function. Myosin light chain kinase (MLCK) is believed to phosphorylate the regulatory light chains of myosin, and thus bring about its contraction. These contractions of myosin and thus the actomyosin cytoskeleton, by virtue of its interaction with tight junction proteins help regulate the permeability of tight junctions.[93] This phosphorylation is in turn regulated by a calcium and ATP dependent mechanism.

c-AMP also affects the permeability of tight junctions by several mechanisms. In the blood-testis barrier in the sertoli cells, it causes proteasome-sensitive ubiquitination of occludin, whereas it interacts with the cytoskeletal system in gallbladder epithelium. c-AMP also regulates microvascular permeability.

1.6.4 Rho Family GTPases

Ras, Rabs and Rho-GTPases have been thought to be involved in tight junction regulation. Rho family of GTPases have been thought to be involved in the regulation of junction assembly as well as paracellular permeability.[35,49,94,95] Ras effectors have been shown to affect tight junction assembly.[96,97] Raf kinases and AF-6 have been shown to be involved in cell differentiation and biogenesis of epithelial tight junctions.[98] They are also involved in phosphorylation of Myosin light chain (MLC) phosphatase, thus leading to its inactivation.

1.6.5 Rho Effector Proteins

Rho associated kinase (ROCK) mediates regulation of paracellular permeability probably though the actin cytoskeleton or directly by acting on tight junction proteins.[99,100]

1.6.6 Crumbs

Crumbs which is an apical membrane protein in *Drosophila* has been localized close to epithelial tight junctions. Crumbs associated proteins have been shown to be further associated with ZO-1 and ZO-3 proteins and PAR3-PAR6-aPKC complex.[101]

1.6.7 Protein Phosphatases

Protein phosphatases dephosphorylate the protein residues and, thus result in enhanced paracellular permeability.[102] Assembly of tight junctions is prevented by overexpression of the catalytic unit of PP2A, whereas its inhibition by various inhibitors like okadaic acid results in increased phosphorylation and translocation of occludin, claudin-1 and ZO-1 to tight junctions.[103,104] While protein phosphatases have not been demonstrated to have any significant influence on actin cytoskeleton and adherens junctions, they very certainly have a disruptive effect on tight junction integrity and assembly.[103-105] Protein phosphatases have been grouped into three families:

a) PPP (Phosphoprotein phosphatases) family and the PPM (metallo-dependent protein phosphatase) which are the major serine/threonine dephosphorylators.

b) PPP includes PP1, PP2A, PP2B etc.

c) Protein tyrosine phosphatase (PTP).

In tandem, protein phosphorylation and dephosphorylation help maintain a balance in the cell and switch the proteins from phosphorylated to dephosphorylated state and vice versa as required. Disruption in these states can lead to a number of diseases including but not limited to cancer, genetic abnormalities, diabetes and hypertension.[86]

1.6.8 Protein Kinases

Protein kinases transfer the gamma phosphate from ATP to a serine, threonine or tyrosine residue in protein substrates. In humans, serine, threonine and tyrosine phosphorylation is 86.4, 11.8 and 1.8% respectively.[86] Like protein phosphatases, protein kinases too are very important components of tight junction regulatory mechanism. Their specific actions of phosphorylating serine, threonine or tyrosine residues influence the integrity and assembly of tight junctions.[34,89,106,107] Their role on tight junction regulation is intertwined with that of protein phosphatases thus making the regulation of tight junctions a dynamic process.

1.6.8.1 Tyrosine kinases

Tyrosine kinases, as the name indicates, phosphorylate the tyrosine residues in the target proteins. Several of such kinases are located in the vicinity of intercellular junctional complex. C-Src and c-Yes are two such protein tyrosine kinases that contribute to the regulation of tight junctions.[108-110] The importance of tyrosine kinases in tight junction regulation stems from the fact that occludin is known to be phosphorylated on tyrosine residues when tight junctions are disrupted.[111] Occludin, when phosphorylated on Tyrosine residues is no longer able to maintain its interaction with ZO-1 and ZO-3.[112,113] There is quite a significant amount of evidence to point towards the role of

14

c-Src in regulation of tight junctions. A study has indicated that activation of c-Src mediates the disruption of tight junctions caused by oxidative stress.[112] It has also been shown that tyrosine kinase inhibitors protect the tight junctions against the oxidative stress-induced disruption in Caco-2 cells. Further studies have shown that over-expression of dominant-negative c-Src delays the disruption of tight junctions when subjected to oxidative stress. This over-expression of kinase-inactive c-Src also accelerates the calcium-induced assembly of tight junctions in Caco-2 cells.[110,114,115] In all these studies, it was observed that phosphorylation of tight junction proteins occludin and ZO-1, and adherens junction proteins β-catenin and E-cadherin on tyrosine residues resulted in their dissociation from the actin cytoskeleton and thus disruption of tight junctions.

1.6.8.2 Serine threonine kinases

These kinases specifically phosphorylate the serine and threonine residues. Protein kinase C activation has been shown to decrease[34,107,116] or increase[106,117,118] paracellular permeability. This suggests a complex signaling pathway at the junctional complex. Protein kinase C(PKC)η[107], PI 3-kinase[119] and MAP kinases[120] are the few prominent serine threonine kinases known to regulate the tight junction structure and function. Several studies have demonstrated the involvement of serine threonine kinases in tight junction regulation. One study has proven that Calphostin C- a specific inhibitor of protein kinase C inhibits the biogenesis of tight junctions.[34] Another study has implicated PI 3-kinase in disruption of tight junctions produced by oxidative stress. It has also been shown that oxidative stress leads to increased association of occludin with PI 3-kinase.[119]

1.7 Role of Protein Phopshorylation/Dephosphorylation in Maintenance of Tight Junctions

Protein phosphorylation/dephosphorylation plays a vital role in all cellular mechanisms. Signaling pathways regulating metabolism, transcription, cell-cycle progression, differentiation, cytoskeletal regulation, apoptosis and intercellular communications are affected by protein phosphorylation/dephosphorylation. A complimentary interplay between various protein kinases and phosphatases is responsible for these functions. Several signaling proteins and molecules modulate the activity of these kinases and phosphatases.[86]

Any interference in this delicate balance between these two groups of enzymes can significantly and adversely affect the cellular processes of cell growth, development, differentiation and migration, and the overall health of the cells. Any interruption in the function of these regulatory enzymes can lead to various diseases and abnormalities like cancer, hypertension, diabetes, cardiac hypertrophy and genetic defects etc. Although our knowledge about the kinases and phosphatases involved in the regulation of

Thr-phosphorylation of occludin is incomplete, PKC, MAPK, PP2A and PP1 are known to be associated with TJ protein complex and regulate the integrity of TJs.[34,107,114]

Occludin is a major component of the tight junction proteins. It is also found in both the detergent-soluble (cytoplasmic) and detergent-insoluble (actin cytoskeleton-associated) fractions of cellular proteins.[121] On SDS-PAGE analysis, in addition to the regular occludin bands, presence of some higher molecular weight bands points towards the possibility that these higher molecular weight bands corresponding to the phosphorylated form of occludin.[111,121,122] Disappearance of these bands when the samples were treated with phosphatases has helped confirm this belief. Further studies have also demonstrated that the sites of phosphorylation in the phosphorylated form of occludin were serine and threonine residues.[123]

Occludin phosphorylation on serine, threonine or tyrosine residues significantly alters the dynamics of the tight junction structure and function. Several studies have drawn a connection between the phosphorylation state of occludin and the integrity of the tight junctions. One study has demonstrated that compared to the cells grown in regular calcium containing medium, those cultured in a low-calcium medium exhibited disruption and disassembly of tight junctions. Further analysis revealed that in the cells incubated in low-calcium medium, occludin was present predominantly in its unphosphorylated form and was localized in the detergent-soluble fraction. These changes were, interestingly, reversed when the regular calcium containing medium was added to the cells, i.e. the junctional integrity and assembly was restored, occludin was seen to be present in its phosphorylated form and redistributed to the detergent insoluble fraction. Immunofluorescence studies have also substantiated this evidence.[111,121] This data indicated that occludin is phosphorylated on serine and threonine residues in intact tight junctions, whereas the disrupted tight junctions are associated with dephosphorylation of occludin on serine and threonine residues.

Phosphorylation of occludin is now a widely known and well established process involved in regulation of tight junction integrity. There is some evidence that points to the role of PKC and ERK in the phosphorylation of occludin on serine and threonine. It is also known that the phosphorylation sites of occludin are located in the ZO-1 binding domain, and some others are present in domains that help target occludin to the tight junction complex.[63]

These studies have established beyond doubt that phosphorylation of occludin on serine and threonine plays a vital role in the regulation of tight junctions. In the present study, we have endeavored to understand the role of this process during TJ assembly and disruption. PKC and MAPK are among the several serine threonine kinases catalyzing this process and regulating the dynamics of the tight junction integrity. Our concentration is to look at the role of PKCζ in the regulation of tight junctions.

1.8 Protein Kinase C in Cell Functions

The protein kinase C (PKC) family consists of 10 serine/threonine kinases. These enzymes are important in regulation of cell proliferation, survival and cell death. PKCs have been classified into three categories:

1. **Classical PKCs**: The classical PKCs include α, β and γ, and are characterized by activation with diacylglycerol (DAG), phosphoserine and phorbol esters. But classical PKCs require calcium for activation.

2. **Novel PKCs**: The novel PKCs include δ, ε, η and θ. These too are activated by diacylglycerol (DAG), phosphoserine and phorbol esters and they do not require calcium for activation.

3. **Atypical PKCs**: They include PKCζ and λ/ι. Like the novel PKCs, atypical PKCs also do not require calcium for activation. Their uniqueness lies in the fact they are not activated by phorbol esters either.[124]

Phorbol esters have been shown to disrupt tight junctions.[106,117,125] However, inhibition of PKCs blocks both tight junction formation and disruption. Transient activation of PKCs is seemingly required for tight junction regulation. Different PKCs have been localized close to tight junctions but other than atypical PKCs nothing much is known about the role of classical and novel PKCs.[34] The atypical PKCs form a complex with PAR3-PAR6 and are important in tight junction regulation and establishment of cell polarity.[126] The PKCs have been thought to be able to affect cell regulation based on environmental stimuli. A change in PKC expression/activity therefore can disrupt the regular homeostatic state of the body and lead to a number of diseases.

1.9 Involvement of an Atypical PKC in Regulation of Occludin Phosphorylation

Calphostin C (PKC inhibitor) has been shown to delay the tight junction assembly in a dose dependent manner. Also, the membrane associated PKC activity has been shown to increase significantly during tight junction assembly. These studies point to the role of PKC in the formation of tight junctions.[34]

During tight junction assembly occludin has been shown to be translocated from the cytoplasm to the region of cell-cell contacts. Treatment with PKC activators has been shown to increase this redistribution of occludin in a dose and time dependent manner. When cells that were undergoing reassembly were lysed and triton soluble and insoluble fractions were made, it was observed that the amount of triton insoluble fraction of occludin increased as the reassembly progressed.[62] This increase was seen to be even more when the assembling cells were treated with a PKC activator. Also, it was shown that this occludin fraction was phosphorylated and the soluble fraction was dephosphorylated. PKCζ, which is an atypical PKC has been shown to colocalize with

ZO-1 at the tight junction.[103,107,116] So it can be suggested that PKCζ might be involved in the phosphorylation of occludin and its localization at tight junctions. In the following chapters, we shall try to find out more about the role of PKCζ in regulation of tight junctions.

CHAPTER 2: MATERIALS AND METHODS

2.1 Cell Culture

Caco-2 and MDCK cell monolayes were used as models of epithelial lining in our studies to define the role of PKCζ in regulation of epithelial tight junctions.

2.1.1 Caco-2 Cell Line

We used Caco-2 cells in several of other studies in our laboratory. This cell line is derived from human epithelial colorectal adenocarcinoma cells.[127] However in culture, they differentiate spontaneously into polarized intestinal columnar cells (enetroctytes) forming a monolayer morphologically and functionally similar to enterocytes in small intestine. The cells forming this cultured monolayer possess an apical brush border and tight junctions between adjacent cells similar to intestinal enterocytes.[128,129] Hidalgo et al showed that Caco-2 cells underwent the above mentioned process of enterocytic differentiation even when they were cultured on polycarbonate transwell inserts. They also found that these monolayers form a polarized epithelial cell monolayer and exhibited uptake and permeability properties similar to small intestine, thus confirming their suitability as a candidate for intestinal epithelial model.[127] It has since been established that Caco-2 cell line is an excellent model for studying intestinal absorption. Artursson et al have demonstrated a correlation between in vitro (Caco-2 cell monolayers) apparent permeability and in vivo (small intestine) absorption of drugs, thus indicating suitability of Caco-2 cell monolayers as model for studying absorption from the intestine.[130]

Caco-2 cells were grown in 100 mm petri dishes or 75 cm^2 flasks. Dulbelco's modified DMEM containing 10% (v/v) fetal bovine serum plus L-glutamine, penicillin, streptomycin and gentamicin was used for cell culture. Cells were kept at 37°C in cell culture incubators with 5% CO_2. Medium was changed every other day. Once confluent, the cells were passaged 1:3 into new dishes and flasks using 0.05% Trypsin/0.53 mM EDTA in HBSS. To maintain sterility, the medium was filtered through 0.22 micron filters and all procedures were done in laminar flow hood.

2.1.2 MDCK Cell Line

MDCK (Madin-Darby Canine Kidney) cell line is developed from normal canine kidney. Earlier used for viral expression studies, it was characterized as a high throughput model for membrane permeability by Grove et al.[131] They showed that permeability characteristics of MDCK cell monolayers grown on polycarbonate transwell inserts mimic those of Caco-2 cells. It gave us a chance to compare our findings over different cell lines from two organ systems (intestine, kidney) and two species (human, dog).

MDCK cell were cultured following the same cell culture protocol as Caco-2 cells mentioned above. However, the fetal bovine serum in growth medium was replaced with serum supreme.

2.2 Transwell Inserts

In order to study the tight junctions, the cells in epithelial monolayers have to be differentiated and polarized with a distinctive apical and basal side. As stated earlier, cells grown on transwell inserts provide an excellent model of a polarized cell monolayer for epithelial transport studies. The transwell inserts (**Figure 2.**1) consist of a porous polycarbonate membrane which is anchored on the base of a cylindrical well. These inserts have a pore size of 0.4 micron and membrane diameter of 6.5 mm, 12 mm and 24 mm. Transwell inserts were purchased from Corning Costar, MA.

2.3 Inhibition of PKCζ by PKCζ-Pseudosubstrate (PS)

2.3.1 Mechanism of Action

PKCζ consists of four functional domains, one of which is the pseudosubstrate domain (**Figure 2.2**). The pseudosubstrate domain blocks the substrate binding cavity of the kinase domain. We purchased myristoylated PKCζ-PS from Genscript corporation with the sequence- Myr-SIYRRGARRWRKL. This was used to inhibit the PKCζ activity in the MDCK and Caco-2 cell monolayers. Activation of PKCζ is dependent upon release

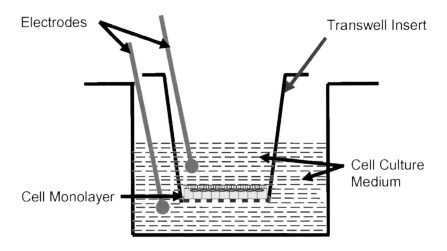

Figure 2.1: A schematic representation of a transwell insert.

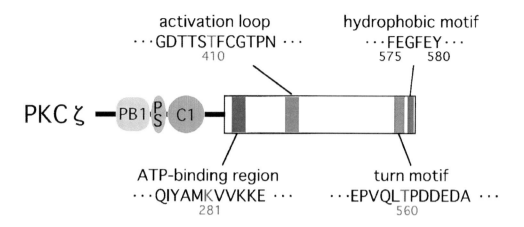

Figure 2.2: Schematic representation of the domain structure of PKCζ.
Reprinted with permission. Chida K, Hirai T. Protein kinase C zeta (PKCζ): activation mechanisms and cellular functions. J Biochem (Tokyo) 2003;133(3):395.[151]

of kinase domain via phosphorylation of Thr-410, which further autophosphorylates Thr-560[124] (**Figure 2.3**).

2.3.2 Stock Preparation

PKCζ PS (MW 1428) comes in powder form. This was reconstituted into a 5mM stock solution in 1% DMSO in distilled water. To preserve the activity it was stored in silicon coated microtubes at -20°C.

2.3.3 Dilution

For use in our experiments, we diluted the stock solution of PKCζ-PS with DMEM 100x to achieve a concentration of 50 μM. Further dilution was done to get lower concentrations, when required.

2.3.4 Inhibition of PKCζ in Caco-2/MDCK Cell Monolayers Grown in Transwell Inserts

The cell monolayers grown on transwell plates were brought to room temperature and baseline Transepithelial Electrical Resistance (TER) was measured to assess the health/viability of cell monolayers. The cell monolayers were washed with serum and glutamine free DMEM 1x and incubated with the same for 1 hour for equilibration. 10 μM, 25 μM or 50 μM PKCζ-PS was added to the apical compartments of transwell inserts. The plates were incubated at 37°C for a maximum of three hours.

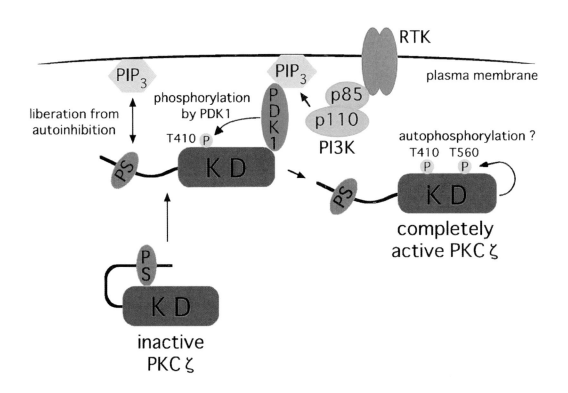

Figure 2.3: Schematic representation of PIP3 and PDK1 in PKCζ activation.
Reprinted with permission. Chida K, Hirai T. Protein kinase C zeta (PKCζ): activation
mechanisms and cellular functions. J Biochem (Tokyo) 2003;133(3):395.[151]

2.4 Measurement of Barrier Function

We evaluated the barrier function by measuring TER (Transepithelial electrical resistance) and paracellular flux of FITC-inulin.

2.4.1 Transepithelial Electrical Resistance (TER) Measurement

The paracellular and transcellular pathways across a cell monolayer are in a parallel circuit. Even though TER is a measure of resistance across both transcellular and paracellular pathways, variations in TER reflect paracellular permeability or more specifically tight junction permeability since tight junctions are the major barrier at the paracellular pathway.[4,6,127,132]

TER measurement was done using an electric voltohmmeter (Millicell-ERS 2 epithelial volt-ohm meter- **Figure 2.4**). It is a battery operated portable apparatus consisting of main unit and a set of electrodes that are connected to the main unit. The main unit has a display which can show either the resistance or potential difference across the membrane. A push button acts as the switch to turn on the apparatus. The 'chopstick' electrodes consist of silver/silver chloride pellet and one electrode is longer than the other so that they can be placed at right angles to the membrane in the apical and basal compartments of the transwell system. (**Figure 2.5**) When the current is passed through the transwell membrane, the meter measures the electric resistance and/or potential difference across the membrane.

To measure the TER, electrodes were sterilized by soaking them in 70% ethanol for 15 min. Then to equilibrate the electrodes they were soaked in DMEM (medium used for experiment) for 15 min. Blank resistance readings were taken using transwell inserts without cultured cells. Since the reading is done with incubating medium and no cell monolayers, it gives us the resistance values of polycarbonate membranes. The transwell plates intended for measurement of TER were allowed to reach room temperature and then resistance was measured by using the apparatus. The measurements were obtained in ohms. Blank TER values for membranes were subtracted from recorded TER values to obtain the actual TER of the cell monolayers. TER readings were converted to 'resistance per unit area' by dividing the actual resistance values by surface area of the membrane (0.33 cm^2, 1.13 cm^2 and 4.52 cm^2 for 6.5 mm, 12 mm and 24 mm transwells respectively). The unit area resistance provides a better comparison between different plate sizes.

2.4.2 FITC-Inulin Flux Measurement

TER is a quick and easy way to measure the epithelial permeability and barrier function in cell monolayers. However, it is not a dynamic measure of the permeability changes. Transepithelial inulin flux is a more dynamic indicator of the permeability changes. To measure the inulin flux across the membrane FITC (Fluorescein

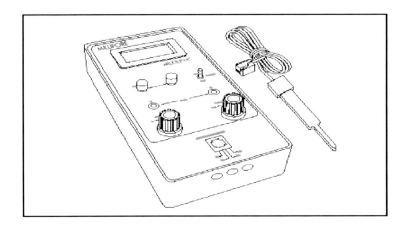

Figure 2.4: Millicell-ERS electrical resistance system.
Millicell is a trademark of Millipore Corporation. The illustration is courtesy of Millipore Corporation. Reprinted with permission. *Millicell*[R]*–ERS User Guide* P17304, Rev. C, 2 (2007). Millipore Corporation Billerica, MA.[152]

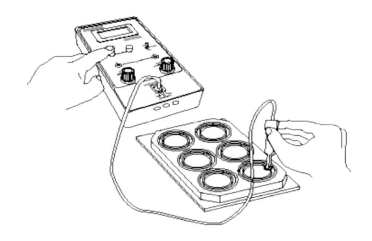

Figure 2.5: Measurement of transepithelial electrical resistance.
Millicell is a trademark of Millipore Corporation. The illustration is courtesy of Millipore Corporation. Reprinted with permission. *Millicell*[R] *ERS User Guide* P17304, Rev. C, 7 (2007). Millipore Corporation Billerica, MA.[153]

isothiocyanate)-inulin was used.[133] It has been shown that tight junctions have pores that allow for passage of ions and solutes across the epithelium. The pore sizes have been estimated to be in the order of 3.5-4 A°.[72] Inulin with a radius of 10 A° thus is a suitable marker to study epithelial permeability. FITC is used as a fluorescent tracer that is tagged to inulin so that it can be read on a fluorescence reader.

50 mg/ml solution of FITC-inulin was placed in the apical compartment to obtain a final dilution of 1:100, and the plates were incubated at $37^\circ C$. 100 μl of medium was taken from the apical as well as basal compartments and transferred to a 96 well fluorescence reader plate. 100 μl incubating medium was used as blank control. The fluorescence was read at excitation of 485 nm and emission of 538 nm using FLx800 microplate fluorescence reader (BioTEK instruments , Winooski, VT) with KC junior software. Net flux as a result of inulin diffusion from the apical to basal compartment was calculated as follows:

$$\frac{Basal\ reading}{Apical\ reading} \times \frac{V2}{V1} \times \frac{100}{T} \times \frac{1}{A} = \%\ flux/hr/cm^2$$

where

    V1 = Volume of medium in the apical compartment
    V2 = Volume of medium in the basal compartment
    T = Time duration of experiment (in hours)
    A = Area of cell monolayer (in cm^2)

2.5 Determination of Cell Viability

To rule out cell damage/apoptosis by PKCζ-PS viability studies- Lactate dehydrogenase (LDH) assay and WST assay were done on Caco-2/MDCK cell monolayers in transwells after treatment with PKCζ-PS.

2.5.1 Lactate Dehydrogenase (LDH) Assay

LDH assay is based on the fact that LDH (a cytoplasmic enzyme) is released from cells when the membrane is damaged. It is to be noted that though there are other enzymes such as phospahtases released upon cell injury, LDH is a stable enzyme and quantification of LDH release has been widely used as a marker of cell injury. LDH is an oxidoreducatse which causes a coupling reaction with the substrate present in the reagent. Upon addition of the reagent, LDH oxidizes lactate to pyruvate which then reacts with tetrazolium salt INT to form formazan. (**Figure 2.6**) The formazan dye is water soluble and can be read in spectrophotometer at 490nm. The amount of formazan formed is proportional to the number of dead/damaged cells.[134,135]

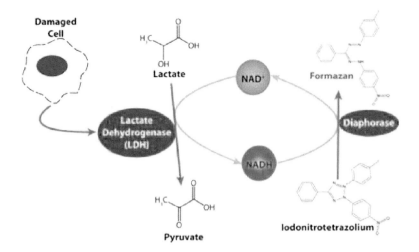

Figure 2.6: The LDH Assay.
Reprinted with permission.The LDH Assay. Retrieved from
http://www.gbiosciences.com/CytoscanLDHCytotoxicityAssayKit.aspx. Accessed on 06
May 2011.[154]

2.5.2 WST Assay

Unlike LDH assay which measures cell death/damage, the WST assay measures
the metabolic activity of cells. The stable tetrazolium salt WST-1 is cleaved to a soluble
formazan dye by a complex cellular mechanism at the cell surface. This reaction is
dependent on the production of NAD(P)H by glycolysis in viable cells. Therefore, the
amount of formazan dye formed is proportional to the number of metabolically active
(viable) cells in the culture.[136]

2.6 Tight Junction Assembly by Calcium Switch

Calcium is known to be involved in the assembly of adherens junctions.[137]
Calcium depletion has been shown to break the adherens as well as the tight junctions in
cell monolayers. The breakdown of tight junctions is not a direct result of calcium
depletion. It is rather a consequence of adherens junction breakdown. This is a reversible
process upto a point and calcium replenishment can lead to denovo assembly of tight
junctions in cell monolayers.[138] This knowledge gave researchers a great tool to study the
reassembly process and signaling mechanisms associated with it. This is known as the
'calcium switch' model. For our study, we looked at TJ reassembly after calcium
replenishment with and without inhibiting PKCζ.

2.6.1 TJ Reassembly in Caco-2 Cells Using a Calcium Chelator (EGTA)

Caco-2 cells were grown on transwell inserts until confluent i.e. 10, 12 or 14 days for 6.5 mm, 12 mm or 24 mm transwell inserts respectively. Basal TER measurement was done to assess the health of monolayers and the cells were washed 1X with serum and glutamine free medium. The cells were incubated at $37^{o}C$ for 1 hour with serum/glutamine free medium for equilibration. The plates were then allowed to come to room temperature by keeping them outside for 5 min. TER measurement was done and then calcium depletion was caused using 4 mM EGTA. Tight junction integrity was monitored by measuring TER and observing the cell monolayers under microscope. A decline in TER and cell rounding indicated the disruption of junctions. When the TER values reached 30% of basal TER (usually 25-30 min), the process is stopped by washing the cells twice with regular medium containing calcium to completely remove EGTA. Regular medium with calcium was added to the transwells, and monolayers were incubated at $37^{o}C$. The denovo assembly of tight junctions in the monolayer was monitored by TER and inulin flux measurements every 30 minutes.

2.6.2 Calcium Switch with Low Calcium Medium in MDCK Cells

Low calcium medium (LCM) was used to induce TJ breakdown in MDCK cells. LCM was constituted by adding calcium to commercially available calcium free DMEM so as to attain a final concentration of 2 μM. Similar protocol as for Caco-2 cells was followed for calcium switch except that the cells were incubated with LCM for 14-16 hr. Also the cells are not washed with regular medium to stop the process, the LCM is just replaced with regular medium.

2.7 Immunofluorescence Staining and Confocal Microscopy of Tight Junction Proteins

Caco-2/MDCK cells were grown to confluence on polycarbonate membranes in transwell inserts and after the required treatment(s), cells were fixed and stained for tight junction proteins.

2.7.1 Fixing the Cells

Fixation immobilizes antigens while retaining cellular architecture. Caco-2/MDCK cells were fixed at scheduled time points or at the end of the experiment, using either acetone-methanol or paraformaldehyde. The cell monolayers were washed twice using ice cold Phosphate buffered saline (PBS) containing NaCl, KCl, Na_2HPO_4, KH_2PO_4 in distilled water at pH 7.2 to stop the reaction.

2.7.1.1 Fixing with acetone-methanol

Cold acetone-methanol in 1:1 concentration was prepared a day before and stored in a glass bottle at -70ºC. While the cells were being washed with PBS the acetone-methanol is brought to the experiment hood using temperature resistant gloves. After washing with ice cold PBS, the cells were bathed in cold acetone: methanol for 5 minutes, air dried for 5 minutes at room temperature and then stored at -20°C.

2.7.1.2 Fixing with 3% paraformaldehyde

Fresh 3% paraformaldehyde in PBS was prepared on the day of the experiment. After washing with cold PBS, cells were treated with 3% paraformaldehyde for 15 minutes at room temperature. Then, paraformaldehyde was removed by washing with cold PBS 3 times for 10 minutes each and fixed cell monolayers were stored at 4ºC in PBS containing 0.05% sodium azide.

2.7.2 Cutting and Trimming of the Membrane

A small piece of membrane is cut from the stored membrane and immediately immersed in 1X PBS in a 24 well plate. Any rough edges were trimmed with fine scissors as these would cause problems during mounting stage.

2.7.3 Washing of the Membrane

Paraformaldehyde fixed membranes were washed with PBS 3 times for 10 minutes each. Acetone-methanol fixed cells were rehydrated twice with PBS for 10 minutes each.

2.7.4 Permeabilization of the Membrane

Membrane permeabilization allows for better penetration of antibodies into the cells. The cells were permeablilized by treating with 1 ml of 0.2% TritonX-100 (prepared in PBS) for 5 minutes. After permeabilization membranes are washed three times for 10 minutes each with cold PBS. Care has to be taken though that there is not overpermeabilization as that leads to protein disruption and affect staining.

2.7.5 Blocking

Blocking of cell monolayers prevents non specific binding of IgG. The cell monolayer was blocked with 4% milk in TBST (20mM Tris, pH 7.2, 150mM NaCl, Tween20) for 30 minutes at room temperature.

2.7.6 Incubation with Primary Antibodies

The cell monolayers were incubated with the primary antibodies in 150μl of 4% milk containing Mouse anti-occludin monoclonal antibody (3μg/ml) and Rabbit anti-ZO-1 polyclonal antibody (4μg/ml), or Mouse anti-E-cadherin monoclonal antibody (3μg/ml) and Rabbit anti-β-catenin polyclonal antibody. (4μg/ml). The cell monolayers were treated with primary antibodies in a humidifying chamber in the dark for one hour.

2.7.7 Incubation with Secondary Antibodies

The membranes were washed three times with 1% milk for 10 minutes each. They were then incubated with the secondary antibody depending on the primary antibodies used in the previous step. We used anti-mouse IgG conjugated with Alexafluor 488 (1:100) and anti-rabbit IgG conjugated with Cy3 (1:100) in 4% milk. The cells were incubated with the secondary antibody in a humidifying chamber in the dark for one hour and then washed with 1x PBS three times for 10 minutes each.

2.7.8 Mounting of Membranes on Slides

The membranes were mounted on glass slides by placing a small drop of the mounting fluid (DABCO + glycerol) on the slide and placing the membrane on the slide with the apical surface facing up. Cover slips (size zero) were carefully placed on the membrane and then sealed using nail paint on the edges of the cover slip. The slides were labeled and kept in the dark at 4°C for imaging later on.

2.7.9 Observation and Processing

The fluorescence was examined using a confocal laser-scanning microscope (Zeiss LSM510 PASCAL) as a series of images from 1μm XY sections. Iris and gain were adjusted according to the intensity of the stain but same settings were used to compare all samples in each experiment. Images were stacked using the Image J software (National Institute of Health) and processed by Adobe Photoshop (Adobe Systems, San Jose, CA).

2.8 Preparation of Whole Cell Extracts

After the stipulated treatments, cell monolayers were washed twice with cold PBS (NaCl 137mM, KCl 2.7 mM, Na_2HPO_4 10mM, KH_2PO_4 1.76 mM at pH7.4) and lysates were prepared using heated lysis buffer-D (SDS 0.3% v/w, Tris 10 mM at pH 7.4 with sodium vanadate 10 μM, sodium fluoride 100 μM and protease inhibitor cocktail 10μl/ml). After extraction the samples were homogenized by sonication. Protein estimation was done as mentioned below and 3x Laemmli's sample buffer equal to half the volume of the sample was added. The mixture was heated at 100°C for 10 minutes and stored at -20°C.

2.9 Preparation of Detergent-Soluble and Detergent-Insoluble Cell Fractions

Actin cytoskeletal fraction is known to be associated with plasma membrane. The transmembrane proteins are anchored by the actin cytoskeletal fraction and on lysis with detergent remain in the insoluble fraction.[122] After completion of treatment of cell monolayers with different reagents as required by the study, monolayers were washed twice with cold 1x PBS. Lysis buffer-CS (20mM Tris buffer containing 1.0% Triton X-100, 2 µg/ml leupeptin, 10 µg/ml aprotinin, 10 µg/ml bestatin, 10 µg/ml pepstatin-A, 2 mM vanadate, and 1 mM PMSF) was used to lyse cells. The lysate was centrifuged at 15,600 g for 5 min at 4°C and the supernatant was transferred to labeled tubes. The resultant, detergent insoluble, pellet contains the actin cytoskeletal fraction and the supernatant contains the detergent soluble fraction of cell proteins. The pellet was suspended in 200 µl of Lysis buffer CS and homogenized with sonicator. Aliquots were taken from both fractions for protein assay. The protein samples were mixed with half the volume of 3x Laemmli's sample buffer and heated at 100°C for 10 minutes and stored at -20°C.

2.10 Protein Estimation

Protein concentration in the lysates was estimated using BCA protein estimation method. The protein estimation kit has two reagents A and B; these were mixed in a ratio of 1:40 to get the reagent mix. 5µl samples of cell extracts were taken in triplicates and put in a 96 well plate. BSA protein standards (100-1000 µg/ml) were used in duplicates to obtain a reference protein value curve. 200 µl of reagent mix was added to the samples and standards. The plate was incubated at 37°C for 10 minutes and then read at 562 nm using SPECTRAMAX 190 plate reader (Molecular devices, CA). The values obtained were put in Microsoft Excel software and using the standard concentration curve plotted, protein concentrations were calculated.

2.11 Immunoprecipitation of Tight Junction Proteins

Caco-2 or MDCK cell monolayers were washed with cold 1x PBS and the proteins were extracted in Lysis buffer-CS or Lysis buffer-D as mentioned above. After protein estimation, 300 µg of protein was aliquoted and mixed with equal volume of 2x immunoprecipitation buffer in 1.5 ml microfuge tubes. The samples were incubated for 16-18 hours on a rocker at 4°C with 2 µg rabbit polyclonal anti-occludin, mouse monoclonal anti-PP2A, rabbit polyclonal anti-PP1 or rabbit polyclonal anti-phosphoserine or anti-phosphothreonine antibody. 25 µl washed protein-A sepharose beads were added and samples were put back on the rocker for one hour. The supernatant was discarded and beads were washed with 1x immunoprecipitation buffer three times.

30

25 µl of 2x Laemmli's sample buffer was added and the samples heated at 100°C for 10 min. Tubes were spun down at 10,000 rpm and the supernatant was used for immunoblot analysis.

2.12 Immunoblotting

The previously prepared samples of whole cell extract or detergent soluble and insoluble fractions or immunoprecipitated proteins stored in Laemmli's sample buffer were heated at 100°C for 10min. 30 µg of protein/sample was loaded on Nupage-novex 7% tris acetate pre-cast gels (Invitrogen, Carlsbad, CA) and run in MOPS-SDS gel running buffer at 120 V for about 90 minutes till the dye migrated to the lower end of the gel. This leads to elecrophoretic separation of proteins based on their molecular weight. The smaller proteins move faster than bigger proteins and thus can be identified on a film. A colorimetric molecular weight marker is also run with the samples.

PVDF membranes were charged with methanol for 45 sec and the gels were then transferred to PVDF membranes. The transfer was done at 4°C at 100V for 90 minutes using transfer buffer. To ensure there is no overheating of the transfer buffer, the transfer apparatus was kept in an ice chamber. The membranes were taken out of the transfer chamber and blocked with 5% milk (or 5% BSA for phospho-antibodies) in TBST (Tris base, NaCl and TWEEN 20 dissolved in distilled water, pH adjusted to 8.0) for 1 hour at room temperature. The membrane was probed with various primary antibodies in 3% milk (or 3% BSA for phopsho-antibodies) in TBST - rabbit anti-phosphoserine (1:3000), rabbit anti-phosphothreonine (1:3000), mouse anti-occludin (1:3000), rabbit anti-ZO-1 (1:3000), rabbit anti-PKCζ (1:3000), mouse anti-claudin-4 (1:2000) and rabbit anti-claudin-3 (1:2000) antibodies for 12-16 hours at 4°C on a rocker. The membrane was washed 5 times with 1x TBST on the rocker at room temperature for 5 minutes each. After that, the membrane was probed with HRP-conjugated anti-mouse or anti-rabbit secondary antibodies (1:10,000) prepared in 3% milk or 3% BSA in TBST for 1 hour at room temperature on a rocker. The membrane was washed again 5 times with 1x TBST.

The blot was developed using the ECL chemiluminescence method using ECL solutions 1 and 2 (Amersham, Arlington Heights, IL). ECL reagents are added to each other just before developing and spread as thin uniform film on the membrane. The secondary antibody binds with the primary antibody and the ECL (containing peroxide) acts as a substrate for the horse-radish peroxidase which catalyses the oxidation of luminol in the presence of peroxide. This oxidation-reduction reaction oxidizes luminal to an oxidized product which is in an excited state and leads to emission of light as it decays back to the ground state over time. Radiography films can record this emitted light and on developing show us the various proteins detected depending on their molecular weight. For re-probing when needed, the membranes were incubated with 1x stripping buffer (Thermo scientific 21059) for 30 min. on a rocker. After washing off the stripping buffer twice with TBST they were blocked and re-probed as described above.

2.13 Recombinant Proteins

Recombinant C terminal domain of human occludin (C-terminal 150 amino acids) was produced as GST fusion protein (GST-Ocl-C) in E. coli BL21DE cells and purified using GSH-agarose. cDNA for the C-terminal tail of human occludin (amino acids 378-522) was amplified using the cDNA clone for human occludin (kind gift from Dr. Van Italie, University of North Carolina, Chapell Hill, NC) and inserted into pGEX2T vector. The sequence of interest inserted into pGEX2T vector was transformed in highly competent DH5α cells. The transformed cells were grown in Lurea Broth medium and purified using GSH-agarose. Point mutations of T400, T403, T404, T424, and T438 to Ala were introduced in wild-type GST-Occludin-C nucleotide sequence, using Stratagene Quik Change II Site Directed Mutagenesis Kit (Agilent Technologies, Santa Clara, CA) as per manufacturer suggested protocol, and they were expressed as above.

2.14 Pairwise Binding Assay

To determine the direct interaction between occludin and PKCζ, GST-Ocl-C (10 μg) was incubated with recombinant, purified PKCζ. GST (10 μg) was used as control. Both GST and GST-Ocl-C were incubated with 30, 100, 300 and 500 ng of pure recombinant PKCζ in binding buffer (PBS containing 0.2% Triton X100, 1 mM vanadate, and 10 mM sodium fluoride) for 3 hours at 37°C on an inverter. GSH agarose beads were washed with binding buffer three times. GST/GST-Ocl-C was pulled down by binding to 30 μl of 50% GSH-agarose slurry at 37°C for 1 hour. The amounts of PKCζ bound to GSH-agarose pull down were determined by immunoblot analysis. The blots were probed for PKCζ as well as GST for control.

2.15 Occludin Phosphorylation In Vitro

To look at role of PKCζ in occludin phosphorylation, in vitro phosphorylation of occludin C-terminus by PKCζ was examined. GST-Occludin-C (10 μg) was incubated with 500 ng of active PKCζ in 20 mM MOPS, pH 7.2, containing 25 mM β-glycerophosphate, 2.25 mM MgCl2, 0.2 mM ATP and 1 mM dithiothreitol. Following 3 hour incubation at 30°C reaction mixture was immunoblotted for p-Thr, PKCζ and GST.

2.16 Design and Synthesis of Antisense Oligos

For designing antisense oligonucleotides, sequence of PKCζ (Genbank accession number NM_002744) was obtained from the NCBI human database in the Pubmed archives using a BLAST search program. The sequence obtained from the database was double-checked and another BLAST search was performed to rule out any similar sequences in the human genome in the NCBI human database. The search yielded no similar results, which confirmed the unique sequence of PKCζ. We designed two antisense oligonucleotides using the PKCζ sequence. The nucleotide sequences of PKCζ were then compared to other PKC isoforms, using the CLUSTAL W program, which aligned the nucleotide sequences of PKCs to enable us to pinpoint areas in the nucleotide sequences of the enzyme, which were unique and did not match the sequences in the other isoforms. To double check these chosen nucleotide sequence areas, we ran a thorough blast search of the known human genome databases and found no matches other than PKCζ from which these sequences had been chosen, and thus, confirmed the uniqueness of these nucleotide sequences. We then went about designing the reverse complement or the antisense strands to the sequences that we had isolated and so came up with the following antisense oligonucleotides for PKCζ. We also designed a scrambled antisense to serve as an adequate control for our experimental setup.

| Sense PKCζ 1 | 1331-1350 | CAAGCTCACAGACTACGGCA |
| Antisense PKCζ 1 | | TGCCGTAGTCTGTGAGCTTG |

| Sense PKCζ 2 | 1641-1660 | AATAAGGACCCCAAAGAGAG |
| Antisense PKCζ 2 | | CTCTCTTTGGGGTCCTTATT |

2.17 Transfection of Caco-2 and MDCK Cells with Antisense Oligos

Caco-2 or MDCK cells were seeded on 6 well cluster plates. Cells were grown until 60-70% confluent. Transfection solutions A and B were made. Solution A was made by incubating 3µl of Oligofectamine with 15µl of Optimem for 10 min at room temperature. Solution B was made by adding 1 µg scrambled or antisense oligonucleotide specific for PKCζ in 4 µl Plus Reagent and 175 µl Optimem. Solution B was added to solution A and incubated for 20 min at room temperature. The cells in cluster plate were washed twice with Hank's balanced salt solution (HBSS), and transfection mix was added to the wells. The plates were incubated at 37°C for 4-6 hrs and after this time, 1 ml of cell culture medium containing fetal bovine serum was added to each well followed by overnight incubation at 37°C. Next morning, the cell monolayers were trypsinized and seeded in 6.5 mm transwell plates. Experiments were conducted starting 3 days after transfection. Reduction in the expression of PKCζ was verified by immunoblot analysis of whole cell lysates prepared from transfected cells.

2.18 Construction of Expression Vector for PKCζ shRNA

A vector-based short hairpin RNA (shRNA) method was used to silence gene expression of human PKCζ. Two targeting sequences were chosen against the nucleotide sequence of human PKCζ gene (GenBank No. NM_002477) using the Dharmacon web site [Target1: GAATCGTGAGGATCGTATA (nucleotide position, 498-516), Target2: AGAAGTTCCTTCAGTACAA (2993-3011)].

The sequences were further verified by BLAST search on the known human genome databases, and no matches were found other than PKCζ, confirming the uniqueness of these sequences. To construct the shRNA vectors, two pairs of oligonucleotides containing the antisense sequence, a hairpin loop region (TTGATATCCG) and the sense sequence with cohesive BamHI and HindIII sites were synthesized (Sigma Genosys, St Louis, MO) as follows:

Top strand1:
5'GATCCCGTATACGATCCTCACGATTCTTGATATCCGGAATCGTGAGGATCGT ATATTTTTTCCAAA-3'
and bottom strand1:
5'-AGCTTTTGGAAAAAATATACGATCCTCAGATTCCGGATATCAAGAATCGTG AGGATCGTATACGG-3';
top strand2:
5'-GATCCCGTTGTACTGAAGGAACTTCTTTGATATCCGAGAAGTTCCTTCAGT ACAATTTTTTCCAAA-3'
and bottom strand2:
5'-GCTTTTGGAAAAAATTGTACTGAAGGAACTTCTCGGATATCAAAGAAGTTC CTTCAGTACAACGG-3'.

The top and bottom strands were annealed and cloned into BamHI and HindIII sites of the pRNAtin-H1.2 vector (**Figure 2.7**, GenScript Corp., Piscataway, NJ) (pR vector), which induces expression of shRNA by H1.2 promoter and cGFP protein by CMV promoter. Successful insertion of the shRNA constructs into the vector was confirmed by releasing the oligonucleotides by digesting with BamHI and HindIII and sequencing.

2.19 Statistical Analyses

Student's *t*-test was used to compare the observed data in the two different groups. Significance in all tests was set at 95% or greater confidence level.

2.20 Transfection of Caco-2 and MDCK Cells with Expression Vectors

MDCK cells were seeded on 6 well plates a day before transfection. The cells were transfected, using 1 ml antibiotics-free DMEM containing 10% FBS, 1 µg DNA plasmid (Empty vector, AS1 or AS2), 1 µl Plus reagent, and 3 µl Lipofectamine-LTX for each well. After 20 hours, the cell monolayers were trypsinized and seeded in 6.5mm transwell plates. Reduction in PKCzeta protein expression was verified by immunoblot analysis.

2.21 Ex Vivo Studies

To make this study more physiologically relevant the effects of PKCζ inhibition on mouse ileum were studied. Ileal sections extracted from adult mice were used to study

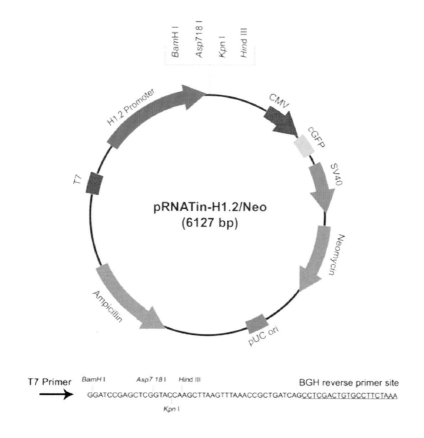

Figure 2.7: pRNATin-H1.2/Neo vector.

Reprinted with permission from Genscript Corporation. pRNATin-H1.2/Neo vector. Retrieved from http://www.genscript.com/product_001/marker/code/SD1223/ siRNA%20Expression%20Vector/pRNATin_H1_2_Neo/SD1223.html. Accessed on 06 May 2011.[155]

the effects of PKCζ inhibition on the TJ structure *ex-vivo*. The experimental protocol was approved by Institutional animal care and use committee (IACUC) and department of comparative medicine (DCM). C57BL/6J mice (*Mus musculus*) were ordered from The Jackson Laboratory, Bar Harbor, Maine. Fig. After the required quarantine, they were transferred to the main animal house facility. Here, they were bred (following IACUC guidelines) and after a stable colony was obtained, 8-12 weeks old adult mice were used for experiments. C57BL/6J mice were used as they are easy to breed and robust.

The mice were transferred to new clean cages (one each) in the morning and transferred from animal facility to the experiment room. Here they were weighed and one mouse at a time was anesthetized using isoflurane obtained from the department of comparative medicine following IACUC guidelines. The mice were dissected to remove the intestine by laparotomy. Extracted intestines were placed in isotonic saline solution. Mice were euthanized by cervical dislocation and the cadaver was placed in biohazard bags. The ileum was flushed with saline to remove undigested food and it was cut open longitudinally.

Thereafter, 1 cm long ileal segments were cut out using a pair of sharp scissors and incubated with varying concentrations of PKCζ pseudosubstrate for 3 hours. Post-treatment, Ileal sections were washed with saline and mucosa was scraped using a glass slide and glass plate. The mucosal scrapings were transferred to cold lysis buffer CS at 4°C and kept on ice. Detergent insoluble and soluble fractions were prepared as described above. Protein estimation was done on the resulting protein fractions, and gel electrophoresis was done. Immunoblotting was done for TJ proteins (Occludin, ZO-1, E-Cadherin and β-catenin) as described above.

Another set of ileal sections, after treatment with PKCζ pseudosubstrate were placed in OCT and frozen with liquid nitrogen for cryosectioning. 10 μ thick sections were prepared on labeled glass slides (Department of Pathology, U.T.H.S.C.). Slides were stored at -70°C until staining.

The slides were fixed in 35 ml of acetone-methanol solution and immunostained for various TJ proteins as mentioned above.

CHAPTER 3: RESULTS

3.1 Inhibition of PKCζ Activity Disrupts TJs in Caco-2 Cell Monolayers

3.1.1 Rationale

Previous studies have indicated the involvement of a PKC in the maintenance of tight junctions via occludin phosphorylation. It is also known that PKCζ (an atypical PKC) is localized at the junctions and was thought to be one of the possible candidates for occludin phosphorylation. Our goal was to check if PKCζ was the PKC involved in tight junction regulation. To study this, we inhibited PKCζ in the cells by using PKCζ pseudosubstrate which is a myristolyated short peptide with the sequence Myr-SIYRRGARRWRKL.

3.1.2 Inhibition of PKCζ by PKCζ Pseudosubstrate

As mentioned before, PKCζ activation depends upon the release of substrate binding site by autophosphorylation. The pseudosubstrate binds to the substrate binding site on PKCζ and prevents its activation. Caco-2 cells were grown in transwell inserts as described in 'Materials and Methods'. PKCζ-PS was used to inhibit PKCζ activity in cell monolayers as mentioned before. Barrier function was evaluated by measuring TER and Inulin flux every hour for three hours. The dose response of PKCζ-PS was assessed by treating the cell monolayers with varying concentrations (10, 25, 50 and 100 μM) of PKCζ-PS and measuring TER and Inulin flux 3 hours after incubation. To rule out if the inhibition of PKCζ causes cell death/damage, we performed LDH assay and WST assay in cell monolayers before and after treatment with PKCζ-PS for three hours.

3.1.3 Inhibition of PKCζ Leads to Time-Dependent Disruption of TJ Barrier Function in Caco-2 Cell Monolayers

We observed that compared to controls, cell monolayers treated with PKCζ-PS exhibited a progressive and significant decline in TER over a three hour time period (**Figure 3.1A**). The cell monolayers treated with PKCζ-PS also demonstrated a similar time-dependent and significant increase in paracellular flux of FITC-inulin, as compared to control monolayers (**Figure 3.1B**).

3.1.4 PKCζ-PS Produces a Dose-Dependent Disruption of Barrier Function in Caco-2 Cell Monolayers

Upon treating the cell monolayers with varying concentrations of PKCζ-PS, we observed that as compared to control monolayers, inhibition of PKCζ produced a

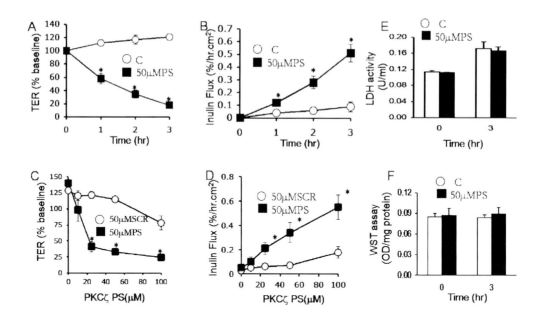

Figure 3.1: Inhibition of PKCζ activity disrupts TJs in Caco-2 cell monolayers.
A-D: Caco-2 cell monolayers were incubated with or without different doses for 3 hours
(C, D) of PKCζ-PS or for varying times at 50 μM PKCζ-PS (A, B). Barrier function was
evaluated by measuring TER (A, C) and unidirectional flux of FITC-inulin (B, D).
Values are mean ± s.e.m. (n = 6). Asterisks indicate the values that are significantly
(P < 0.05) different from corresponding control values. E, F: The cell viability was
assessed by measuring LDH activity in the incubation medium (E) or mitochondrial
activity in the cell by WST assay (F) at 3 hour after PKCζ-PS administration.

disruption of barrier function as evidenced by a decrease in TER (**Figure 3.1C**) and increase in paracellular flux of FITC-inulin (**Figure 3.1D**). This disruptive effect of PKCζ inhibition on barrier function in Caco-2 cell monolayers was found to be directly proportional to dose of the PKCζ-PS administered.

3.1.5 PKCζ-PS Treatment Does Not Affect the Cell Viability

In order to make sure that these changes in functional integrity of tight junctions as described above were not due to cell death, we performed cell viability assays i.e. LDH assay and WST assay on the cell monolayers before and after subjecting them to treatment with 50 μM PKCζ-PS. We observed that PKCζ-PS treated cell monolayers did not show any significant change in LDH activity as compared to control monolayers (**Figure 3.1E**). Similarly, WST assay also did not demonstrate any significant difference in mitochondrial activity between the control cell monolayers and those treated with PKCζ-PS (**Figure 3.1F**).

3.2 Inhibition of PKCζ Activity Leads to Redistribution of TJs in Caco-2 Cell Monolayers

3.2.1 Rationale

After demonstrating the effect of inhibition of PKCζ on the functional integrity of tight junctions in Caco-2 cell monolayers, we wanted to look at the effect of inhibition of PKCζ on localization and distribution of TJ proteins.

3.2.2 Inhibition of PKCζ By PKCζ-PS Adversely Affects the Tight Junction Integrity in Caco-2 Cell Monolayers

To determine the effect of inhibition of PKCζ on localization of TJ proteins, cell monolayers treated with PKCζ-PS were fixed and stained for Occludin and ZO-1. Immunostaining of cell monolayers and confocal microscopy showed that compared with controls, the cell monolayers treated with PKCζ-PS had a reduced junctional localization of TJ proteins Occludin and ZO-1, which were found to be translocated to the intracellular compartment in response to inhibition of PKCζ (**Figure 3.2**).

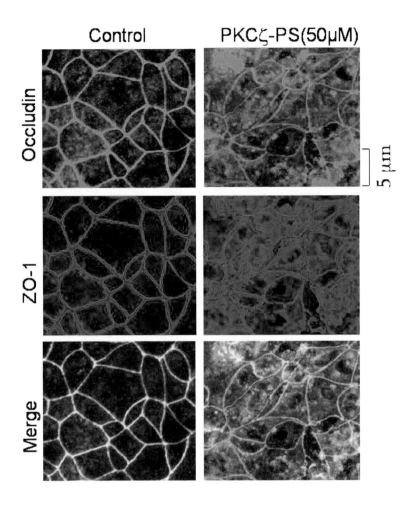

Figure 3.2: Inhibition of PKCζ activity leads to redistribution of TJs in Caco-2 cell monolayers.
Caco-2 cell monolayers incubated with or without 50 μM PKCζ-PS for 2 hr were fixed and stained for occludin and ZO-1 by immunofluorescence method. Images were collected by confocal microscopy.

3.3 PKCζ -PS Disrupts the Adherens Junction in Caco-2 Cell Monolayers

3.3.1 Rationale

Previous studies have shown that inhibition of PKCη causes TJ disruption while sparing adherens junctions. Since PKCη is a novel PKC whereas PKCζ is an atypical PKC, we were interested in seeing whether PKCζ had any effect on adherens junctions in Caco-2 cell monolayers. This study was aimed at studying the role of PKCζ in the regulation of adherens junctions.

3.3.2 Inhibition of PKCζ Leads to Adherens Junction Disruption in Caco-2 Cells

Confluent Caco-2 cell monolayers grown in transwell inserts were incubated with 25 or 50 μM of PKCζ-PS for 3 hours. The monolayers were fixed in acetone-methanol and stained for E-cadherin and β-catenin by immunofluorescense technique. Confocal microscopy of the stained slides demonstrated that treatment of cell monolayers with PKCζ-PS resulted in redistribution adherens junction proteins E-cadherin and β-catenin from the intercellular junctions into the intracellular compartment. We further observed that this redistribution of E-cadherin and β-catenin was much stronger in cell monolayers treated with 50 μM of PKCζ-PS compared with those treated with 25 μM of PKCζ-PS (**Figure 3.3**).

3.4 Inhibition of PKCζ Activity Disrupts TJs in MDCK Cells

3.4.1 Rationale

To determine the general role of PKCζ in TJ regulation in different epithelia, we also evaluated the effect of PKCζ-PS on the maintenance of TJ integrity and the de novo assembly of TJs in MDCK cell monolayers. Caco-2 cells being a model of intestinal epithelial cells and MDCK cells representing renal epithelium, their use in our studies could give us an idea regarding the function of PKCζ in different organ systems.

3.4.2 Inhibition of PKCζ Activity by PKCζ-PS

MDCK cell monolayers were grown in transwell inserts and were used for experiments after reaching confluence. In order to study the time-response of PKCζ-PS in MDCK cell monolayers, cells were treated with 50 μM PKCζ-PS, and its effect was studied on paracellular permeability of FITC-inulin at hourly intervals over a three-hour time period. We also studied the dose-response PKCζ-PS, MDCK cell monolayers were treated with 10, 25, 50 or 100 μM of PKCζ-PS. Inulin flux was measured at the end of three hours. The cell monolayers were fixed and stained for TJ proteins as mentioned

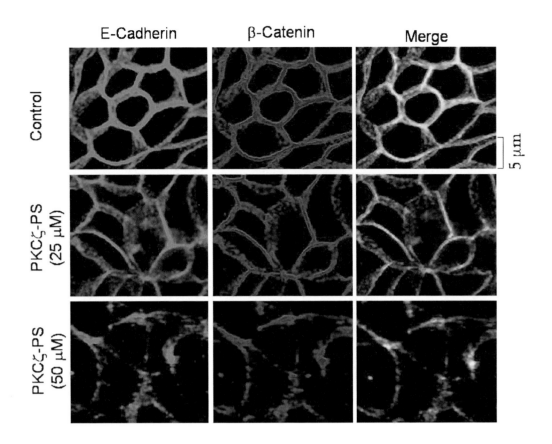

Figure 3.3: Inhibition of PKCζ activity disrupts adherens junctions in Caco-2 cell monolayers.
Caco-2 cell monolayers were incubated with or without PKCζ-PS for 2 hours. Cell monolayers were fixed and stained for E-cadherin and β-catenin by immunofluorescence method. Images collected by confocal microscopy.

before. Viability studies were done to rule out cell death/damage due to treatment with PKCζ-PS.

3.4.3 PKCζ Inhibition Leads to Time-Dependent and Dose-Dependent Increase in Inulin Flux in MDCK Cell Monolayers

Administration of 50 μM PKCζ-PS increased inulin permeability in MDCK cell monolayers. This increase in permeability was observed to increase with time over a three-hour period. (**Figure 3.4A**). We also observed that after 3 hours of PKCζ-PS treatment, treatment with 50 μM PKCζ-PS resulted in greater increase in inulinpermeability compared with the use of 25 μM PKCζ-PS, thus indicating a dose dependent effect of PKCζ-PS on tight junction integrity (**Figure 3.4B**).

3.4.4 Determination of Cell Viability

PKCζ-PS treated cell monolayers did not show a significant difference in LDH release (**Figure 3.4C**) or mitochondrial activity (**Figure 3.4D**) as compared to control monolayers over a 5 hour period.

3.4.5 PKCζ-PS Caused TJ Disruption and Internalization of TJ Proteins in MDCK Cell Monolayers

Inhibition of PKCζ by PKCζ-PS led to loss of junctional structure and internalization of tight junction proteins occludin and ZO-1 away from the junction into the intracellular compartment compared to control cell monolayers (**Figure 3.4E**).

3.5 Inhibition of PKCζ Activity Delays Calcium-Induced Assembly of TJs in Caco-2 and Cell Monolayers

3.5.1 Rationale

Having established the role of PKCζ in the maintenance of tight junctions, we now wanted to study the role of PKCζ in the de novo assembly of tight junctions. So we evaluated the effect of inhibition of PKCζ on the assembly of tight junctions using the calcium switch model.

3.5.2 Calcium Switch Experiment

Fully confluent Caco-2 and MDCK cell monolayers grown on transwell inserts were used for this set of experiments. EGTA (to attain a final concentration of 4 mM)

43

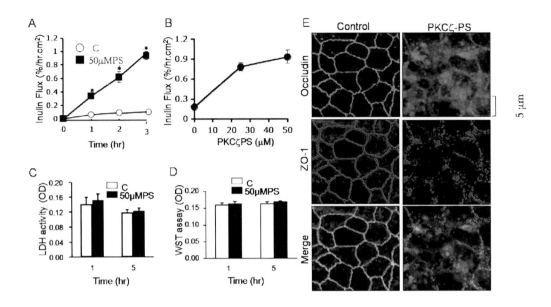

Figure 3.4: Inhibition of PKCζ activity disrupts barrier function and delays TJ assembly in MDCK cell monolayers.

A, B: MDCK cell monolayers were incubated with or without different doses for 3 hours (B) of PKCζ-PS for varying times 50 μM PKCζ-PS (A). Barrier function was evaluated by measuring unidirectional flux of FITC-inulin. Values are mean ± s.e.m. (n = 6). Asterisks indicate the values that are significantly (P < 0.05) different from corresponding control values. C, D: The cell viability assessed by measuring LDH activity in the incubation medium (C) or mitochondrial activity in the cell by WST assay (D) at 3 hour after PKCζ-PS administration. Values are mean ± sem (n = 6). E: Caco-2 cell monolayers incubated with or without 50 μM PKCζ-PS for 2 hr were fixed and stained for occludin and ZO-1 by immunofluorescence method. Images collected by confocal microscopy.

was used to disrupt junctions in Caco-2 cells and the reassembly was initiated by adding regular calcium-containing medium into the transwells. For MDCK cells low calcium medium was used. We monitored the functional integrity of cell monolayers during disruption as well as the reassembly process by measuring TER and inulin flux every hour. The monolayers were fixed in either acetone-methanol or paraformaldehyde after four hours and used for immunostaining.

3.5.3 PKCζ-PS Inhibits the Calcium-Induced Recovery in TER and Inulin Flux in Caco-2 Cell Monolayers

Treatment of cell monolayers with EGTA led to a rapid decline of TER (**Figure 3.5A**) and the increase of inulin flux within 30 minutes (**Figure 3.5B**). Replacement of calcium gradually increased TER and decreased inulin permeability. However, calcium-induced increase in TER and decrease in inulin flux were significantly attenuated in the presence of PKCζ-PS.

3.5.4 PKCζ-PS Attenuates Calcium-Induced Reassembly of TJ in MDCK Cells

During the calcium switch experiment, cells incubated with PKCζ-PS showed a consistently high inulin flux even after calcium-induced reassembly as compared to control monolayers which exhibited a consistent and significant decrease in inulin permeability over a three hour period of reassembly (**Figure 3.5C**).

3.6 PKCζ-PS Inhibits the Calcium Induced Relocation of Occludin and ZO-1 to Junctions in Caco-2 Cell Monolayers

EGTA treatment induced redistribution of occludin and ZO-1 from the intercellular junctions into the intracellular compartment (**Figure 3.6**), which were reassembled back at the intercellular junctions by calcium replacement. The presence of PKCζ-PS prevented this calcium-induced reassembly of occludin and ZO-1 at the junctions (**Figure 3.6**).

3.7 Inhibition of PKCζ Disrupts TJs in Mouse Ileum

3.7.1 Rationale

All our studies done so far used either the cell culture model or the recombinant pure proteins. Even though there exists a considerable body of evidence that the results obtained by using cell studies or in vitro studies are very reliable and similar to those seen with in vivo models, we still tried to confirm the physiological relevance of our specific observations using an ex vivo model involving isolated mouse intestinal epithelial tissue.

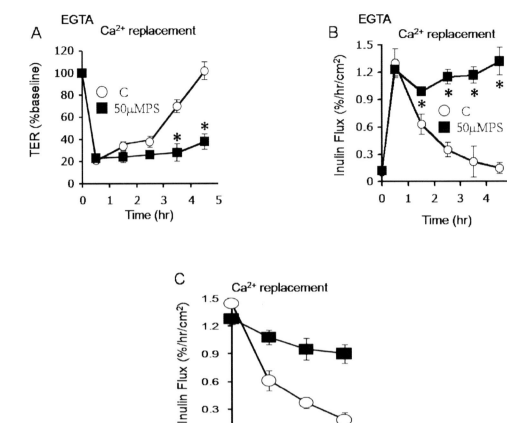

Figure 3.5: PKCζ-PS prevents calcium-induced assembly of TJs in Caco-2 and MDCK cells.
A, B: Caco-2 cell monolayers were treated with 3 mM EGTA for 30 min to deplete extracellular calcium. Regular medium with calcium and with or without PKCζ-PS (10 μM) was then replaced. TER (A) and inulin permeability (B) were measured at varying times. C: MDCK Cell monolayers were incubated in low calcium medium (LCM) for 16 hours to deplete extracellular calcium. Regular medium with high calcium and with or without PKCζ-PS (3 μM) was then replaced. Inulin permeability (C) was measured at varying times after calcium replacement. Values are mean ± sem (n = 6). Asterisks indicate the values that are significantly (P < 0.05) different from corresponding control values.

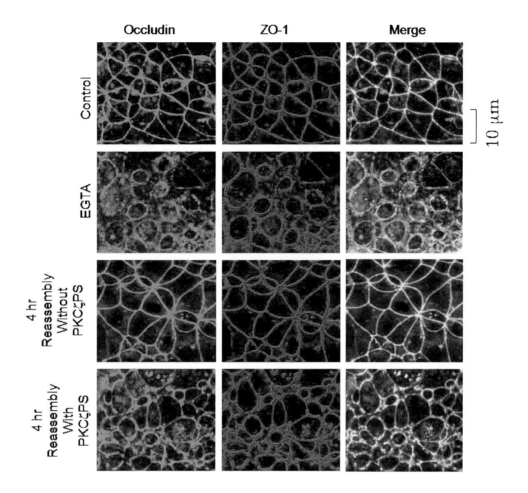

Figure 3.6: PKCζ-PS inhibits the calcium induced relocation of Occludin and ZO-1 to junctions in Caco-2 cell monolayers.
Caco-2 cell monolayers were treated with 3 mM EGTA for 30 min to deplete extracellular calcium. Regular medium with calcium and with or without PKCζ-PS (10 μM) was then replaced. Cell monolayers at various stages of tight junction assembly were fixed and stained for occludin and ZO-1 by immunofluorescence method. Images collected by confocal microscopy.

3.7.2 Inhibition of PKCζ Activity with PKCζ-PS

Short mouse ileal segments were obtained as described in 'Materials and methods'. These segments were incubated with or without varying concentrations (10, 25 or 50 μM) of PKCζ-PS at 37°C for one hour. Mucosa from these ileal segments was scraped and lysed with lysis buffer 'CS'. Detergent-insoluble and soluble fractions of epithelial proteins were prepared as described earlier. These protein fractions were analyzed by SDS-PAGE and immunoblotted for TJ proteins occludin and ZO-1. Another set of ileum segments were preserved in OCT for preparation of frozen sections. The frozen sections were then stained for tight junction proteins occludin and ZO-1, and adherens junction proteins E-cadherin and β-catenin.The immunoblot analysis revealed that compared to control ileum segments, not treated with PKCζ-PS, inhibition of PKCζ by PKCζ-PS resulted in decrease in the levels of Triton-insoluble fractions (**Figure 3.7 A**) of occludin and ZO-1 with a concomitant increase in the levels of Triton soluble fractions (**Figure 3.7 B**) of these proteins suggesting a redistribution of these proteins from the intercellular junctions into the intracellular compartment of the cells. We further noticed that this redistribution increased with the increasing dose of PKCζ-PS used.

Immunostaining and confocal microscopy showed that occludin and ZO-1 are co-localized at the intercellular junctions of ileal epithelium not treated with PKCζ-PS. Treatment with 10 or 25 μM PKCζ-PS resulted in disruption of the junctional organization of occludin and ZO-1 in a dose-dependent manner (**Figure 3.8**). Similar breakdown of junctional organization of E-cadherin and β-catenin was also seen in response to treatment with PKCζ-PS (**Figure 3.9**).

3.8 Reduced Expression of PKCζ with Antisense Oligos Attenuates TJ Integrity in Caco-2 Cell Monolayers

3.8.1 Rationale

In the studies described so far, we have used a pharmacological inhibitor in order to inhibit the effect of PKCζ. While pharmacological agents constitute an important and reliable tool to study the effect of an enzyme, we employed molecular techniques of knockdown of PKCζ using antisense oligos, to corroborate our findings with PKCζ-PS. Reducing the expression of a protein of interest in cell signaling pathway is a more specific and reliable technique than using pharmacological inhibitors.

3.8.2 Transfection of Caco-2 Cells with Antisense Oligonucleotides

We designed two antisense oligos (AS-1 and AS-2) specific to the sequence of human PKCζ as described under 'Materials and Methods'. Caco-2 cells were transfected with antisense oligos specific for PKCζ or missense oligo (MS). We prepared whole cell

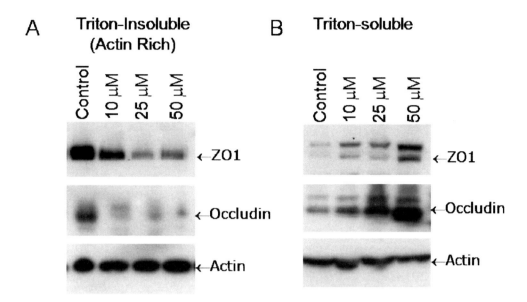

Figure 3.7: Inhibition of PKCζ leads to translocation of TJ proteins from the junction to cytoplasm.
Mouse ileal strips were incubated with or without varying doses of PKCζ-PS for one hour. Triton-insoluble (A) and Triton-soluble (B) fractions were prepared from mucosal scrapings of ileum and immunoblotted for different TJ proteins.

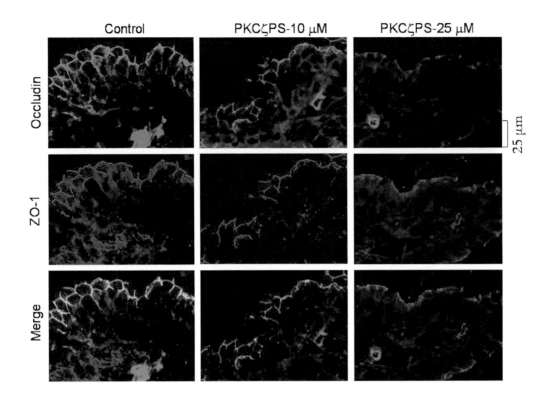

Figure 3.8: PKCζ-PS attenuates TJ integrity in mouse ileum.
Mouse ileal strips were incubated with or without varying doses of PKCζ-PS for one hour. Tissues were cryo-fixed and sections were stained for occludin and ZO-1 by immunofluorescence method.

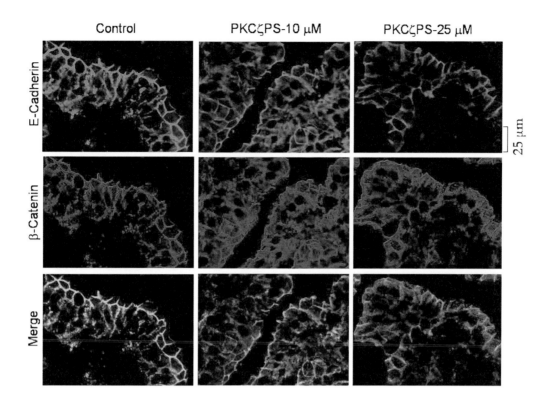

Figure 3.9: PKCζ-PS attenuates AJ integrity in mouse ileum.
Mouse ileal strips were incubated with or without varying doses of PKCζ-PS for one hour. Tissues were cryo-fixed and sections were stained for E-cadherin and β-catenin by immunofluorescence method.

extracts from transfected cells and immunoblotted them for PKCζ in order to confirm the reduction in expression of PKCζ. The extracts were also blotted for PKCλ to rule out the reduction of PKCλ expression by using antisense nucleotide for PKCζ.

3.8.3 Transfection of PKCζ with Antisense Nucleotides Leads to Reduced Expression of PKCζ while PKCλ Expression Is Unaffected

Transfection with AS-1 and AS-2 showed a significant reduction in the levels of PKCζ in Caco-2 cells (**Figure 3.10A**). We found that transfection with AS-2 resulted in much greater reduction in PKCζ expression in Caco-2 cells compared with AS-1. However, PKCλ expression was not affected (**Figure 3.10A**). This was confirmed by densitometric analysis which showed a significant decline in expression levels of PKCζ by both AS-1 and AS-2 whereas PKCλ expression was not significantly affected.

3.8.4 PKCζ Knockdown Leads to TJ Disruption in Caco-2 Cell Monolayers

Transfection with either AS-1 or AS-2 significantly reduced TER (**Figure 3.10C**) and enhanced inulin permeability (**Figure 3.10D**) in Caco-2 cell monolayers. We also observed that this disruption in barrier function was more marked in Caco-2 cell monolayers transfected with AS-2 compared to cells transfected with AS-1. This difference in the barrier function disruption between the two groups of cells was in line with the levels of reduction in expression of PKCζ caused by AS-1 and AS-2.

Immunostaining and confocal microscopy for tight junction proteins Occludin and ZO-1 revealed that knockdown of PKCζ by AS-2 in Caco-2 cells (**Figure 3.10E**) induced redistribution of these proteins from the intercellular junctions into the intracellular compartments, indicating a delayed assembly of TJs with reduced PKCζ expression.

3.9 Reduced Expression of PKCζ with Antisense Oligos Attenuates TJ Integrity in MDCK Cell Monolayers

3.9.1 Rationale

In the studies described so far, we have used a pharmacological inhibitor in order to inhibit the effect of PKCζ. While pharmacological agents constitute an important and reliable tool to study the effect of an enzyme, we employed molecular techniques of knockdown of PKCζ using antisense oligos, to corroborate our findings with PKCζ-PS. Reducing the expression of a protein of interest in cell signaling pathway is a more specific and reliable technique than using pharmacological inhibitors.

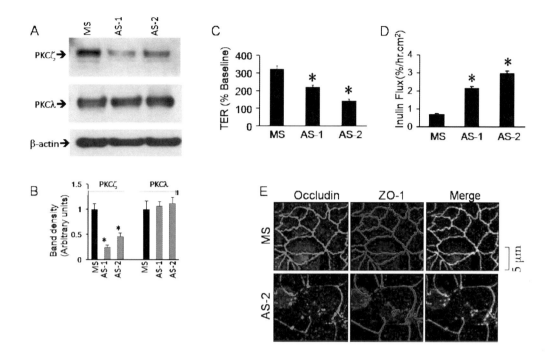

Figure 3.10: Reduced expression of PKCζ by antisense oligos attenuates TJ integrity in Caco-2 cells.

A & B: Caco-2 cells were transfected with missense oligo (MS) or two different antisense oligos (AS-1, AS-2), and the levels of PKCζ and PKCλ were measured by immunoblot analysis. Band densities evaluated by densitometric analysis (B). Values are mean ± sem (n = 6). Asterisks indicate the values that are significantly (P < 0.05) different from corresponding values.

3.9.2 Transfection of Caco-2 Cells with Antisense Oligonucleotides

We designed two antisense oligos (AS-1 and AS-2) specific to the sequence of human PKCζ as described under 'Materials and Methods'. MDCK cells were transfected with antisense oligos specific for PKCζ or missense oligo (MS). We prepared whole cell extracts from transfected cells and immunoblotted them for PKCζ in order to confirm the reduction in expression of PKCζ.

3.9.3 Transfection of PKCζ with Antisense Nucleotides Leads to Reduced Expression of PKCζ while PKCl Expression is Unaffected

Transfection with AS-1 and AS-2 showed a significant reduction in the levels of PKCζ in MDCK cells (**Figure 3.11A**). We found that transfection with AS-1 resulted in much greater reduction in PKCζ expression in Caco-2 cells compared with AS-2.

3.9.4 PKCζ Knockdown Leads to TJ Disruption in Caco-2 and MDCK Cell Monolayers

Transfection with either AS-1 or AS-2 significantly reduced TER (**Figure 3.11B**) and enhanced inulin permeability (**Figure 3.11C**) in MDCK cell monolayers. We also observed that this disruption in barrier function was more marked in MDCK cell monolayers transfected with AS-1 compared to cells transfected with AS-2. This difference in the barrier function disruption between the two groups of cells was in line with the levels of reduction in expression of PKCζ caused by AS-1 and AS-2.

3.10 PKCζ Knockdown by shRNA Leads to TJ Disruption and Attenuation of TJ Assembly in MDCK Cells

3.10.1 Rationale

The studies done so far using the PKCζ-PS and antisense oligonucleotides have clearly indicated that PKCζ plays an important role in regulation of tight junction integrity and the barrier function in both Caco-2 and MDCK cells. In order to further establish the role of PKCζ in regulation of tight junctions in the MDCK cells, we used the shRNA to reduce the expression of PKCζ in these cells. Use of shRNA has some distinct advantages over antisense oligos in that shRNA is a more stable and more specific technique as compared to antisense oligos.

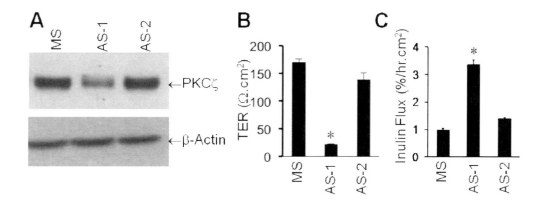

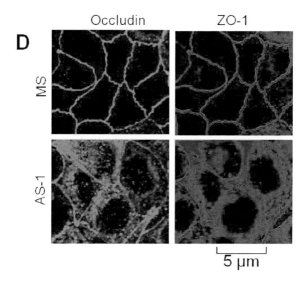

5 μm

Figure 3.11: Reduced expression of PKCζ by antisense oligos attenuates TJ integrity in MDCK cells.
A: MDCK cells were transfected with missense oligo (MS) or two different antisense oligos (AS-1, AS-2), and the levels of PKCζ measured by immunoblot analysis. B& C: After 3 days TER(B) and Inulin flux (C) were measured and plotted. Values are mean ± sem (n = 6).Asterisks indicate the values that are significantly (P < 0.05) different from corresponding values. D: The cell monolayers were fixed and stained for Occludin and ZO-1 and observed under confocal microscope.

3.10.2 Transfection of MDCK Cells with shRNA to PKCζ

PKCζ-specific shRNA was designed and inserted into GFP-tagged pRNATin-H1.2/Neo vector as described in 'Materials and Methods'. MDCK cells were transfected with this shRNA construct using lipofectamine. Cells transfected with the empty vector were used as control. The transfection was confirmed by observing the cells for green fluorescence under fluorescence microscope, as wells as by immunoblotting the cell extracts for PKCζ.

3.10.3 Knockdown of PKCζ by shRNA Leads to TJ Disruption and Attenuation of Calcium Induced Reassembly in MDCK Cells

Immunoblotting of extracts from cells transfected with shRNA showed reduced expression of PKCζ as compared to control cells transfected with the empty vector (**Figure 3.12A**). The shRNA-transfected cell monolayers demonstrated a significantly lower TER compared to vector-transfected cell monolayers (**Figure 3.12B**) indicating that reduced expression of PKCζ leads to delayed development of barrier function in these cells. Similarly, the paracellular permeability of FITC-inulin in cell monolayers with reduced PKCζ expression was significantly greater compared to cells with normal PKCζ expression (**Figure 3.12C**).

Immunofluorescence microscopy showed the presence of GFP in about 35% of cells in both vector and shRNA-transfected cells (**Figure 3.12D**). Co-immunostaining for GFP and ZO-1 showed that ZO-1 was localized at the intercellular junctions in vector-transfected cells, in both GFP-positive and GFP-negative cells. On the other hand, in shRNA-transfected cells, GFP-positive cells showed intracellular localization of ZO-1 (**Figure 3.12D**), whereas in GFP-negative cells ZO-1 was localized predominantly at the intercellular junctions.

We also studied the impact of reduced expression of PKCζ by shRNA on calcium-induced development of barrier function in MDCK cells. We observed that the calcium-induced re-assembly of tight junctions over a three-hour period was significantly delayed in shRNA-transfected MDCK cell monolayers compared to that in vector-transfected cell monolayers (**Figure 3.12E**).

3.11 Inhibition of PKCζ Activity Reduces Detergent Insoluble Fractions of TJ Proteins

3.11.1 Rationale

TJ proteins Occludin and ZO-1 are anchored to the actin cytoskeleton in the intact epithelial monolayer and therefore, these proteins are pulled down along with the actin

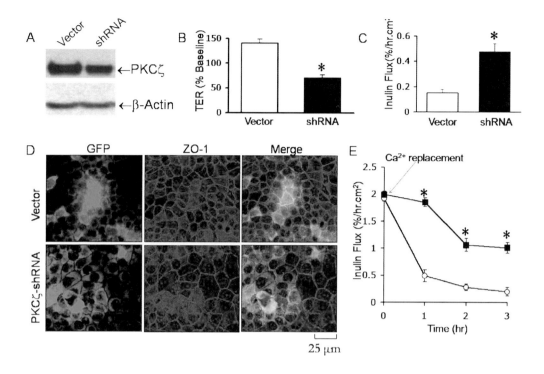

25 μm

Figure 3.12: Reduced expression of PKCζ by shRNA attenuates TJ integrity in MDCK cell monolayers.

A: shRNA specific for PKCζ in pRNAtinH1.2 vector or the empty vector was transfected into MDCK cells. PKCζ expression was determined by immunoblot analysis. B, C: Barrier function was evaluated by measuring TER (B) and inulin permeability (C) on day 3 after seeding. D: Fixed cell monolayers were stained for GFP and ZO-1 (D). Tight junction assembly in transfected cells was evaluated by calcium switch method. E: Inulin permeability measured during calcium-induced reassembly in vector transfected and shRNA-transfected cell monolayers . Values are mean ± sem (n = 6). Asterisks indicate the values that are significantly ($P < 0.05$) different from corresponding values for vector transfected cells.

cytoskeleton (detergent-insoluble fraction) during immunoprecipitation. Previous studies have showed that disruption of TJs is associated with loss of TJ proteins.

3.11.2 PKCζ Inhibition Reduced Detergent Insoluble Fractions of TJ Proteins

Caco-2 cell monolayers were treated with or without with different concentrations (10, 25 or 50 μM)of PKCζ-PS for 30, 60 or 120 minutes. Detergent-insoluble fractions of cellular proteins were prepared in lysis buffer CS, from these cell monolayers at the end of each time point. These detergent-insoluble protein fractions were then subjected to electrophoresis using SDS-PAGE and then immunoblotted for various TJ proteins like ZO-1, ZO-3, Occludin, Claudin-1 and Claudin-3. Immunoblotting was also done for p-PKCζ. We observed that treatment with PKCζ-PS reduced the amounts of ZO-1, ZO-3, occludin and Claudin-1 in the detergent-insoluble fraction of cell proteins in a time and dose-dependent manner (**Figure 3.13A**). We also noticed that the effect of inhibition of PKCζ on Claudin-1 levels was delayed compared to that on ZO-1, ZO-3 and occludin. Interestingly. the level of detergent-insoluble fraction of Claudin-3 was unaffected by PKCζ-PS treatment. Analysis of the active PKCζ (phospho-PKCζ) indicated that PKCζ-PS inactivated PKCζ in a dose-dependent manner. Co-immunoprecipitation studies involving Occludin and ZO-1 in Caco-2 cell monolayers treated with or without PKCζ-PS showed that PKCζ-PS treatment did not alter the co-immunoprecipitation of occludin and ZO-1 (**Figure 3.13B**).

3.12 PKCζ Directly Interacts with the C-Terminal Domain of Occludin

3.12.1 Rationale

Previous study indicated that PKCζ is localized at the vicinity of TJs in MDCK cell monolayers (24). However, the direct interaction of PKCζ with TJ proteins and its involvement in the phosphorylation of TJ proteins is unknown. We conducted the following study to assess if there is a direct interaction of PKCζ with occludin.

3.12.2 PKCζ Directly Interacts with and Binds to Occludin (C-Terminus)

We prepared the C-terminal domain of human occludin as a GST-fusion protein (GST-Ocl- C) and incubated it with varying concentrations (30, 100, 300 and 500 ng) of recombinant PKCζ for 3 hours. Direct binding between GST-Ocl-C and PKCζ was evaluated by GST pull down assay.

We observed a definite interaction between PKCζ and GST-Ocl-C. It was also observed that binding of PKCζ with GST-Ocl-C increased with the increasing concentrations of PKCζ used in the study, thus pointing towards a dose-response relationship **(Figure 3.14)**. However, no significant difference was noticed between the

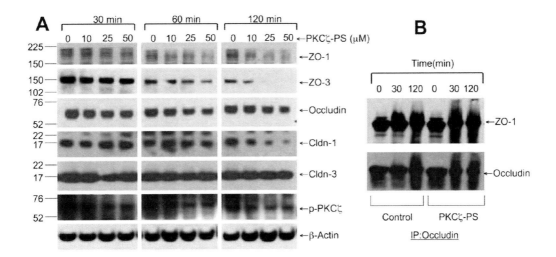

Figure 3.13: PKCζ-PS reduces the levels of detergent-insoluble TJ proteins.
A: Caco-2 cell monolayers were incubated with or without different doses of
PKCζ-PS for varying times. Triton-insoluble fractions were prepared and immunoblotted
for different TJ proteins. B: Occludin was immunoprecipitated from the native protein
extracts from Caco-2 cell monolayers incubated with or without PKCζ-PS for varying
times. Immunocomplexes were then immunoblotted for occludin and ZO-1.

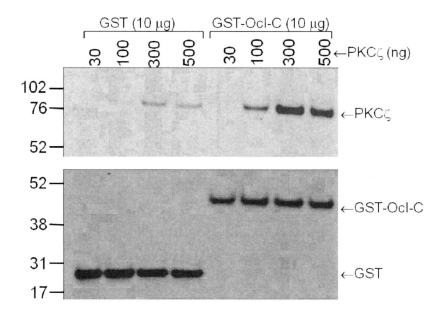

Figure 3.14: PKCζ directly binds to the C-terminal domain of occludin.
Recombinant GST-Ocl-C was incubated with varying amount s of recombinant PKCζ. The GSH-agarose pull down from these samples was then immunoblotted for PKCζ and GST. Values on the left margin of the blots represent the molecular weights of marker proteins.

binding of GST-Ocl-C and 300 ng or 500 ng of PKCζ. We noticed only a negligible and insignificant binding of PKCζ to GST which was used as control in this study. This suggests that PKCζ binds directly to C-terminal tail of occludin, and excludes GST as a binding target for PKCζ.

3.13 PKCζ Activity Is Involved in Ser/Thr phosphorylation of Occludin and ZO-1

3.13.1 Rationale

Occludin is known to be highly phosphorylated on Ser/Thr residues in the intact epithelia, while the disruption of TJ is associated with rapid dephosphorylation of occludin . Since we have established the role of PKCζ in regulation of tight junctions, we now wanted to understand the mechanism behind this regulation. One of the putative mechanisms was that PKCζ influences the tight junction integrity and barrier function by modulating the phosphorylation of occludin on Serine and Threonine residues. So, the following studies were aimed at evaluating the role of PKCζ in the phosphorylation of Serine and/or Threonine residues in tight junction proteins.

3.13.2 Purified Recombinant PKCζ Phosphorylates GST-Ocl-C In Vitro on Serine and Threonine Residues and This Phosphorylation Is Inhibited by PKCζ-PS

GST-Ocl-C was incubated with 500 ng of recombinant PKCζ or PKCη in the presence of ATP for 1 hour. PKCζ-PS in varying concentrations (1, 3 or 10 µM) was also added to the reaction mix. The resulting protein mixtures were analyzed by gel electrophoresis using SDS-PAGE and immunoblotted for p-Thr and p-Ser. We observed that PKCζ induced phosphorylation of GST-Ocl-C on both Thr and Ser residues (**Figure 3.15**). Presence of PKCζ-PS in the assay mixture reduced PKCζ-mediated phosphorylation of GST-Ocl-C in a dose dependent manner. We also noticed that incubation with PKCη also induces Ser/Thr phosphorylation of GST-Ocl-C . However, PKCζ-PS failed to affect this phosphorylation by PKCη, indicating the specificity of PKCζ-PS. Further, PKCζ-PS inhibits autophosphorylation of PKCζ in a dose dependent manner making it a possible mechanism of occludin dephosphorylation.

3.13.3 Inhibition of PKCζ Leads to Reduced Ser/Thr Phosphorylation of Occludin in Caco-2 Cell Monolayers

Caco-2 cell monolayers were grown and incubated with or without 50 µM PKCζ-PS for 30, 60, 120 and 180 minutes. Cell were lysed in lysis buffer 'CS' at the end of each time point and detergent-insoluble and soluble fractions prepared. The detergent-insoluble fraction was immunoprecipitated for p-Ser and p-Thr and immunoblotted for occludin, ZO-1 and Claudin-1. The results indicate that Thr- and

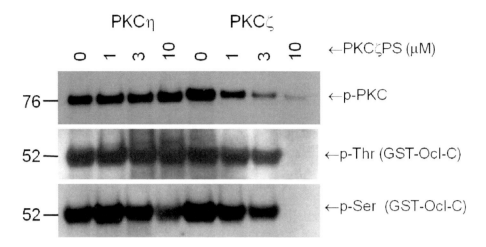

Figure 3.15: In vitro Thr/Ser phosphorylation of GST-OCL-C by recombinant PKCζ is inhibited by PKCζ-PS.
GST-Ocl-C was incubated with PKCζ (500 ng) or PKCη (500 ng) in the presence of varying concentrations of PKCζ-PS and 0.5 mM ATP. Samples were then immunoblotted for p-Thr (A), p-Ser (B) and p-PKC (C).

Ser-phosphorylation of occludin remained unaffected during the incubation time in the absence of PKCζ-PS (**Figure 3.16**). The presence of PKCζ-PS caused a rapid reduction in the phosphorylation of occludin on both Thr and Ser residues. The Ser/Thr phosphorylation of Claudin-1 was, however, unaffected by PKCζ-PS.

3.14 PKCζ Phosphorylates the C-Terminal Tail of Occludin on Specific Thr Residues

3.14.1 Rationale

Once the role of PKCζ in Ser/Thr phosphorylation of occludin was established, our next step was to identify the specific Thr residues that were being phosphorylated by PKCζ. Our analysis of the occludin sequence among various species namely human, dog, mouse, chicken and frog revealed that Thr residues 400, 403, 404, 424 and 438 are conserved residues in occludin C-terminus across these species. In a previous study, PKCη has been shown to phosphorylate occludin C-terminus at Thr 403 and 404. So, we wanted to study which specific Threonine residues are phosphorylated by PKCζ.

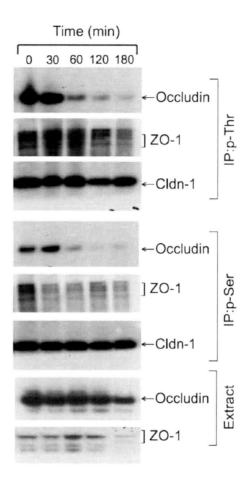

Figure 3.16: PKCζ-PS attenuates Ser/Thr phosphorylation of occludin and ZO-1.
Caco-2 cell monolayers were incubated with or without different doses of PKCζ-PS for varying time periods. p-Thr or p-Ser was immunoprecipitated from the denatured protein extracts. Immunocomplexes were then immunoblotted for different TJ proteins.

3.14.2 Thr Residues 403, 424 and 438 on C-Terminus Occludin Are Phosphorylated by PKCζ

Occludin C-terminus mutants were generated by point mutation and combined with GST to obtain GST-tagged mutant proteins as described in 'Materials and Methods'. 10 μg of each of these Threonine-mutants of GST-tagged C-terminus occludin or wild type GST-Ocl-C were incubated with 500 ng of recombinant PKCζ in the presence of ATP for 1 hour.

Our results indicate that mutation of T438 almost completely attenuated Thr-phosphorylation of GST-Ocl-C whereas mutation of T400 did not significantly influence PKCζ-mediated Thr- phosphorylation of GST-Ocl-C. Mutation of T403, T404 and T424 resulted in partial reduction in theThr-phosphorylation of GST-Ocl-C (**Figure 3.17**).

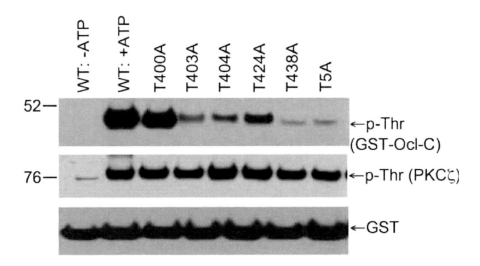

Figure 3.17: C-terminal occludin is phosphorylated on Thr 403, 404 and 438 residues by PKCζ.
Wild type and Thr-mutants of GST-Ocl-C were produced and incubated with PKCζ in the presence of ATP for 3 hours. Phosphorylation was assessed by immunoblot analysis.

CHAPTER 4. DISCUSSION

Epithelial barrier function is an important characteristic that defines multicellular life. The development of barrier function has been deemed an important and necessary evolutionary step in the development of metazoa from single celled organisms.[1] Breakdown of barrier function results in disease pathogenesis.[139,140] Four different types of intercellular junctions are found in vertebrates- tight junctions, desmosomes, gap junctions and adherens junctions.[4] Our focus is to study intestinal barrier function; thus TJs which are the major contributors to barrier function automatically gained importance in our studies. Over the past 30 years major insights have been achieved into the morphology, structure and function of tight junctions. However, TJ regulation is yet to be deciphered fully.[7]

A significant body of evidence indicates that epithelial TJs are regulated by intracellular signaling elements including protein kinases.[87] Protein kinase C has been identified as a possible regulator of barrier function through TJ regulation.[34] Stuart et al have shown that PKC inhibition led to attenuation of TJ assembly and also that PKC activity is increased during TJ assembly. They also showed that an atypical isotype of PKC, the PKCζ localizes at the junctions.[34] The present study provides evidence to the role of PKCζ in the assembly and maintenance of TJs in Caco-2 and MDCK cell monolayers. We showed that PKCζ inhibition led to disruption of barrier function and attenuation of calcium induced reassembly. This study also indicates that PKCζ directly interacts with occludin and induces phosphorylation of occludin on Thr residues.

PKCζ-PS (a cell permeable peptide) has been previously shown to selectively inhibit PKCζ activity in various types of cells. A scrambled peptide was used as control to rule out peptide induced damage to cell monolayers. Measurement of TER (Transepithelial electrical resistance) and FITC-Inulin flux have been shown to be a reliable and convenient method of measuring barrier function of cell monolayers grown on transwell inserts.[127,132] Caco-2 and MDCK cell monolayers have been used as a cheap and easy model to study epithelial permeability. It has been shown that Caco-2 and MDCK cell monolayers grown on transwell inserts mimic intestinal and renal epithelium respectively

The present study showed that administration of PKCζ-PS to Caco-2 and MDCK cell monolayers rapidly disrupted the barrier function as shown by the decrease in TER and increase in inulin permeability in cells treated with PKCζ-PS as compared to control cells. This disruption was shown to be time and dose dependent. Whereas there is a gradual and steady decrease in TER and corresponding increase in Inulin flux in the time course experiment, there is a sharp fall in TER at 25μM of PKCζ-PS in the concentration curve. However, this sharp fall is not mirrored by a corresponding spike in Inulin flux which shows a gradual increase. This can be due to TER being a static indicator of TJ function whereas Inulin flux is a dynamic indicator.

This barrier disruption was not associated with any loss of cell viability as evidenced by no change in LDH release or mitochondrial activity (WST assay) in the cell monolayers after 3 hr incubation with PKCζ-PS. LDH release measures cell death/damage while WST measures the number of viable cells. We do see a change in LDH release after 3 hrs for both PKCζ-PS treated as well as untreated cells but this change can be attributed to cell death damage that occurs due to handling of cells during the experiment.

PKCζ-PS mediated barrier dysfunction was associated with a redistribution of occludin and ZO-1 from the intercellular junctions into the intracellular compartments, indicating that inhibition of PKCζ leads to disruption of TJs. Therefore, inhibition of PKCζ activity in both Caco-2 and MDCK cell monolayers results in disruption of TJs without affecting the cell viability. These results indicate that PKCζ activity is required for the maintenance of TJ integrity once the epithelium is formed. However it has been shown that PKCζ-PS treatment also inhibits PKCλ activity in cells which is another closely related atypical PKC. PKCζ and PKCλ have been shown to have upto 72% similar amino acid sequence identity.[141] Since they both share the conserved PS domain which is the target for PKCζ-PS both are inhibited by treatment with the inhibitor. So maybe both the atypical PKCs are involved in TJ regulation.

This study also looked at the effect of PKCζ-PS treatment on calcium induced reassembly of TJ in Caco-2 and MDCK cell monolayers. The calcium switch model has been used extensively to study the denovo assembly of TJ in cell monolayers. The basic principle as mentioned before is the breakdown of adherens junctions on depletion of calcium. This calcium depletion is attained by either using calcium chelators or low calcium medium. AJ breakdown leads to disruption of TJ which can be reversed by calcium replenishment. This recovery of TJ function can be assessed by measuring TER and Inulin flux. Our results showed that inhibition of PKCζ results in attenuation of the reassembly process in PKCζ-PS treated cell monolayers as compared to control monolayers.

Our study also shows that PKCζ-PS treatment induces redistribution of E-cadherin and β-catenin from the intercellular junctions into the intracellular compartments, indicating that the AJs are disrupted by the inhibition of PKCζ activity. Although AJs do not form a physical barrier to the diffusion of macromolecules across the epithelium, they indirectly regulate the integrity of TJs. Disruption of AJs by calcium depletion is well established to result in disruption of TJs.[142] A recent study has demonstrated that inhibition of PKCη by a specific pseudosubstrate disrupts TJ structure without affecting the AJs.[107] Therefore, PKCζ may regulate the integrity of both TJs and AJs, while PKCη may influence only TJs. Whether the PKCζ-PS mediated disruption of TJs and AJs is sequential or simultaneous is not known.

The role of PKCζ in TJ regulation was further confirmed by the specific knockdown of PKCζ by antisense oligonucleotides, which effectively reduced the level of PKCζ in both Caco-2 and MDCK cell monolayers. To rule out the involvement of PKCλ in the regulation of TJ we designed antisense oligonucleotides that were specific to human PKCζ (AS-1 and AS-2). Both AS-1 and AS-2 were a 100% match to human PKCζ sequence while AS-1 when run against canine PKCζ had a 94% match while AS-2 only 40%.

Antisense oligonucleotides (Both AS-1 and AS-2) to PKCζ reduced TER and elevated inulin permeability in Caco-2 cells, thus demonstrating the disruption of barrier function, which was associated with a redistribution of TJ proteins from the intercellular junctions indicating a delayed assembly of TJ proteins. However, in MDCK cells only AS-1 was effective. This was substantiated by the sequence homology mentioned above.

To confirm the reduced expression of PKCζ, a PKCζ immunoblot was performed which showed a decline in PKCζ expression while PKCλ expression was unchanged. This could be explained by the fact that even though the antisense nucleotides had a 100% sequence homology with PKCζ, it was only 40% with PKCλ. Antisense oligonucleotides to PKCζ reduced TER and elevated inulin permeability, thus demonstrating the disruption of barrier function, which was associated with a redistribution of TJ proteins from the intercellular junctions indicating a delayed assembly of TJ proteins. The advantages of using PKCζ knockdown studies over PKCζ-PS are that it is more specific and also preempts the possibility of PKCζ activation by another mechanism. However it is to be noted that this is transient transfection and thus the studies have to be performed 3 days after transfection which might affect the basal TER and Inulin flux values.

PKCζ-specific shRNA transformed in pRNATin-H1.2/Neo vector, which also contains GFP gene was transfected to MDCK cells. Similar to antisense oligos, shRNA also reduced the barrier function. The expression of GFP allowed us to compare the GFP-positive, transfected cells with the GFP-negative, non-transfected cells in the same monolayer. The results of this study confirmed that junctional distribution of ZO-1 was disrupted only in GFP-positive or shRNA transfected cells, while the junctional distribution of ZO-1 in GFP-negative cells were intact.

A previous study showed that TJ proteins in the intact epithelium are associated with the detergent-insoluble fraction of cells, implicating their tight interaction with the actin cytoskeleton.[51] It was consistently demonstrated that disruption of TJs is associated with a loss of detergent-insoluble fractions of TJ proteins. The present study showed that the PKCζ-PS treatment induces a loss of detergent-insoluble fraction of occludin, ZO-1, ZO-3 and Claudin-1 in a time and dose-dependent manner. The decrease in the level of detergent insoluble ZO-1, ZO-3 and occludin occurred as early as 30 min after PKCζ-PS administration, whereas decrease in detergent-insoluble Claudin-1 was evident only at 120 min of treatment. This indicates that the effect of Claudin-1 may be secondary due to the disruption of TJs by disassembly of occludin, ZO-1, ZO-3 complex.

The level of detergent-insoluble Claudin-3 was unaffected even at 120 min. Interestingly enough, PKCζ-PS did not influence the coimmunoprecipitation of occludin and ZO-1, suggesting that the PKCζ may not influence the occludin-ZO-1 interaction. Loss of TJ integrity in epithelial monolayers by calcium depletion, oxidative stress , acetaldehyde , phorbol esters and pathogen is known to be associated with dephosphorylation of occludin on Ser/Thr residues. However, the precise mechanism involved in this process is unclear. The present study shows that disruption of TJs by the inhibition of PKCζ is also associated with a rapid dephosphorylation of occludin on Ser and Thr residues. Therefore, PKCζ may maintain the TJ integrity by preserving the phosphorylation state of occludin. The present study also shows that PKCζ-PS dephosphorylates ZO-1 on Ser and Thr residues. Altered ZO-1 phosphorylation could also contribute to the disruption of TJs. However, the role of ZO-1 phosphorylation in TJ regulation is not well characterized. Additionally, Ser phosphorylation of occludin and other proteins may also be involved in TJ disruption. Farshori and Kachar have also shown that high concentration of TPA led to decreased phosphorylation of occludin on threonine residues but did not prevent its colocalization with ZO-1.[143] PKC activation by using TPA has been shown to increase paracellular permeability but dephosphorylation of occludin. This led to the hypothesis that there is an intermediary serine/threonine phosphatase that is activated by PKC. Our lab has shown the involvement of PP2A in occludin phosphorylation.[104] PP2A has been shown to be associate with and regulate PKCζ. Therefore, multiple mechanisms may be associated with TJ disruption. Also TPA treatment has been shown to decrease barrier function and increase occludin and ZO-1 transcription. It is to be noted though that the transcription change precedes permeability change.[144] This would suggest that gene expression of tight junction proteins and regulation of tight junction function do not necessarily overlap. Increased expression of PKCα and PKCδ have been correlated with increased paracellular permeability after treatment with phorbol esters. PKCα translocation from the cytosol to the cytoskeleton fraction has been correlated with phorbol ester induced increase in paracellular permeability.[118]

In the present study, we focused our effort on the Thr-phosphorylation of occludin. In order to detect the PKCζ-mediated phosphorylation sites we induced point mutation to T400, T403, T404, T424 and T438, the highly conserved Thr residues in the C-terminal domain of occludin, and prepared the recombinant occludin C-terminal domain as GST-fusion proteins. In vitro phosphorylation by PKCζ showed that PKCζ predominantly phosphorylates T438, T403, T404 and T424. This is somewhat different from the role of PKCη, which phosphorylates T403, T404 and T438, but not T424.[107]

PKCζ has been mainly shown to be regulated by PDK-1 in PI 3-kinase signaling by phosphorylation and activation of PKCζ. PDK-1 has been shown to be associated with PKCζ in vivo. Mutation of Thr 410 site on PKCζ has been shown to block PKCζ phosphorylation by PDK-1. Membrane targeting of PKCζ has been shown to render it constitutively active.[145] Previous study in our lab has shown that MAPK indirectly phosphorylates occludin by phosphorylating PKCζ.[120] That may be one of the putative upstream signals to phosphorylate PKCζ. A previous study in our lab has shown that membrane translocation of PKCε and PKCβ1 is required for EGF mediated protection of TJ from acetaldehyde.[146] However, it was also shown that neither PKCε nor PKCβ1 directly interacts with occludin.[146] Binding of atypical PKC i.e. ζ or λ to Par 3 and Par 6 has been reported in epidermal barrier. Par proteins belong to PDZ family of adaptor proteins. However PKCλ and not PKCζ is the dominant atypical PKC in epidermal tight junctions.[126]

Caco-2 and MDCK cell monolayers have been extensively used to understand the structure and regulation of epithelial TJs, and the information derived from such studies have been extended to animal tissue. The present study shows that incubation with PKCζ-PS disrupts TJs in mouse ileum. PKCζ-PS induces a redistribution of occludin and ZO-1 from the intercellular junctions and reduced the levels of detergent insoluble fractions of occludin and ZO-1. These results demonstrate that PKCζ activity is required for the maintenance of TJ integrity in mouse ileum and confirm the physiologic relevance of the observation made in Caco-2 and MDCK cell monolayers.

This study therefore, demonstrates that PKCζ activity is required for the maintenance of epithelial tight junction. The mechanism of this TJ integrity may involve PKCζ-mediated phosphorylation of occludin and possible other TJ proteins on specific Thr residues.

LIST OF REFERENCES

1.   Cereijido M, Contreras RG, Shoshani L. Cell adhesion, polarity, and epithelia in the dawn of metazoans. Physiol Rev 2004;84(4):1229-1262.

2.   Anderson JM, Van Itallie CM. Tight junctions and the molecular basis for regulation of paracellular permeability. Am J Physiol 1995;269(4 Pt 1):G467-G475.

3.   Lapierre LA. The molecular structure of the tight junction. Adv Drug Deliv Rev 2000;41(3):255-264.

4.   Schneeberger EE, Lynch RD. The tight junction: a multifunctional complex. Am J Physiol Cell Physiol 2004;286(6):C1213-C1228.

5.   Sawada N, Murata M, Kikuchi K et al. Tight junctions and human diseases. Med Electron Microsc 2003;36(3):147-156.

6.   Tsukita S, Furuse M, Itoh M. Multifunctional strands in tight junctions. Nat Rev Mol Cell Biol 2001;2(4):285-293.

7.   Franke WW. Discovering the molecular components of intercellular junctions-a historical view. Cold Spring Harb Perspect Biol 2009;1(3):a003061.

8.   Anderson JM, Van Itallie CM. Tight junctions. Curr Biol 2008;18(20):R941-R943.

9.   Farquhar MG, Palade GE. Junctional complexes in various epithelia. J Cell Biol 1963;17:375-412.

10.   Cereijido M, Gonzalez-Mariscal L, Contreras RG. Epithelial tight junctions. Am Rev Respir Dis 1988;138(6 Pt 2):S17-S21.

11.   Madara JL. Tight junction dynamics: is paracellular transport regulated? Cell 1988;53(4):497-498.

12.   van MG, Simons K. The function of tight junctions in maintaining differences in lipid composition between the apical and the basolateral cell surface domains of MDCK cells. EMBO J 1986;5(7):1455-1464.

13.   Mitic LL, Anderson JM. Molecular architecture of tight junctions. Annu Rev Physiol 1998;60:121-142.

14.   Schneeberger EE, Lynch RD. Structure, function, and regulation of cellular tight junctions. Am J Physiol 1992;262(6 Pt 1):L647-L661.

15. Tsukita S, Tsukita S. [Adherens junction: its structure and function]. Tanpakushitsu Kakusan Koso 1989;34(12 Suppl):1542-1550.

16. Yamada KM, Geiger B. Molecular interactions in cell adhesion complexes. Curr Opin Cell Biol 1997;9(1):76-85.

17. Gumbiner BM. Cell adhesion: the molecular basis of tissue architecture and morphogenesis. Cell 1996;84(3):345-357.

18. Garrod D, Chidgey M. Desmosome structure, composition and function. Biochim Biophys Acta 2008;1778(3):572-587.

19. Garrod D, Chidgey M, North A. Desmosomes: differentiation, development, dynamics and disease. Curr Opin Cell Biol 1996;8(5):670-678.

20. Staehelin LA. Structure and function of intercellular junctions. Int Rev Cytol 1974;39:191-283.

21. Kumar NM, Gilula NB. The gap junction communication channel. Cell 1996;84(3):381-388.

22. Anderson JM. Leaky junctions and cholestasis: a tight correlation. Gastroenterology 1996;110(5):1662-1665.

23. Fanning AS, Mitic LL, Anderson JM. Transmembrane proteins in the tight junction barrier. J Am Soc Nephrol 1999;10(6):1337-1345.

24. Stevenson BR, Anderson JM, Braun ID, Mooseker MS. Phosphorylation of the tight-junction protein ZO-1 in two strains of Madin-Darby canine kidney cells which differ in transepithelial resistance. Biochem J 1989;263(2):597-599.

25. Gonzalez-Mariscal L, Namorado MC, Martin D et al. Tight junction proteins ZO-1, ZO-2, and occludin along isolated renal tubules. Kidney Int 2000;57(6):2386-2402.

26. Stein J, Kottra G. [Intestinal intercellular tight junctions. I. Structure and molecular mechanisms of regulation]. Z Gastroenterol 1997;35(3):205-220.

27. Balda MS, Matter K. Tight junctions at a glance. J Cell Sci 2008;121(Pt 22):3677-3682.

28. Furuse M. Molecular basis of the core structure of tight junctions. Cold Spring Harb Perspect Biol 2010;2(1):a002907.

29. Schneeberger EE, Lynch RD. Tight junctions. Their structure, composition, and function. Circ Res 1984;55(6):723-733.

30. Braga VM, Balda MS. Regulation of cell-cell adhesion. Semin Cell Dev Biol 2004;15(6):631-632.

31. Citi S, Cordenonsi M. Tight junction proteins. Biochim Biophys Acta 1998;1448(1):1-11.

32. Madara JL. Modulation of tight junctional permeability. Adv Drug Deliv Rev 2000;41(3):251-253.

33. Siccardi D, Turner JR, Mrsny RJ. Regulation of intestinal epithelial function: a link between opportunities for macromolecular drug delivery and inflammatory bowel disease. Adv Drug Deliv Rev 2005;57(2):219-235.

34. Stuart RO, Nigam SK. Regulated assembly of tight junctions by protein kinase C. Proc Natl Acad Sci U S A 1995;92(13):6072-6076.

35. Terry S, Nie M, Matter K, Balda MS. Rho signaling and tight junction functions. Physiology (Bethesda) 2010;25(1):16-26.

36. Turner JR. Molecular basis of epithelial barrier regulation: from basic mechanisms to clinical application. Am J Pathol 2006;169(6):1901-1909.

37. Balda MS, Anderson JM, Matter K. The SH3 domain of the tight junction protein ZO-1 binds to a serine protein kinase that phosphorylates a region C-terminal to this domain. FEBS Lett 1996;399(3):326-332.

38. Anderson JM, Stevenson BR, Jesaitis LA, Goodenough DA, Mooseker MS. Characterization of ZO-1, a protein component of the tight junction from mouse liver and Madin-Darby canine kidney cells. J Cell Biol 1988;106(4):1141-1149.

39. Anderson JM, Fanning AS, Lapierre L, Van Itallie CM. Zonula occludens (ZO)-1 and ZO-2: membrane-associated guanylate kinase homologues (MAGuKs) of the tight junction. Biochem Soc Trans 1995;23(3):470-475.

40. Fanning AS, Jameson BJ, Jesaitis LA, Anderson JM. The tight junction protein ZO-1 establishes a link between the transmembrane protein occludin and the actin cytoskeleton. J Biol Chem 1998;273(45):29745-29753.

41. Haskins J, Gu L, Wittchen ES, Hibbard J, Stevenson BR. ZO-3, a novel member of the MAGUK protein family found at the tight junction, interacts with ZO-1 and occludin. J Cell Biol 1998;141(1):199-208.

42. Gonzalez-Mariscal L, Betanzos A, Avila-Flores A. MAGUK proteins: structure and role in the tight junction. Semin Cell Dev Biol 2000;11(4):315-324.

43. Fanning AS, Anderson JM. Zonula occludens-1 and -2 are cytosolic scaffolds that regulate the assembly of cellular junctions. Ann N Y Acad Sci 2009;1165:113-120.

44. Stevenson BR, Heintzelman MB, Anderson JM, Citi S, Mooseker MS. ZO-1 and cingulin: tight junction proteins with distinct identities and localizations. Am J Physiol 1989;257(4 Pt 1):C621-C628.

45. Bazzoni G, Martinez-Estrada OM, Orsenigo F, Cordenonsi M, Citi S, Dejana E. Interaction of junctional adhesion molecule with the tight junction components ZO-1, cingulin, and occludin. J Biol Chem 2000;275(27):20520-20526.

46. Cordenonsi M, D'Atri F, Hammar E et al. Cingulin contains globular and coiled-coil domains and interacts with ZO-1, ZO-2, ZO-3, and myosin. J Cell Biol 1999;147(7):1569-1582.

47. Keon BH, Schafer S, Kuhn C, Grund C, Franke WW. Symplekin, a novel type of tight junction plaque protein. J Cell Biol 1996;134(4):1003-1018.

48. Satoh H, Zhong Y, Isomura H et al. Localization of 7H6 tight junction-associated antigen along the cell border of vascular endothelial cells correlates with paracellular barrier function against ions, large molecules, and cancer cells. Exp Cell Res 1996;222(2):269-274.

49. Nusrat A, Giry M, Turner JR et al. Rho protein regulates tight junctions and perijunctional actin organization in polarized epithelia. Proc Natl Acad Sci U S A 1995;92(23):10629-10633.

50. Wittchen ES, Haskins J, Stevenson BR. Protein interactions at the tight junction. Actin has multiple binding partners, and ZO-1 forms independent complexes with ZO-2 and ZO-3. J Biol Chem 1999;274(49):35179-35185.

51. Nusrat A, Turner JR, Madara JL. Molecular physiology and pathophysiology of tight junctions. IV. Regulation of tight junctions by extracellular stimuli: nutrients, cytokines, and immune cells. Am J Physiol Gastrointest Liver Physiol 2000;279(5):G851-G857.

52. Turner JR, Rill BK, Carlson SL et al. Physiological regulation of epithelial tight junctions is associated with myosin light-chain phosphorylation. Am J Physiol 1997;273(4 Pt 1):C1378-C1385.

53. Liu Y, Nusrat A, Schnell FJ et al. Human junction adhesion molecule regulates tight junction resealing in epithelia. J Cell Sci 2000;113(Pt 13):2363-2374.

54. Furuse M, Fujita K, Hiiragi T, Fujimoto K, Tsukita S. Claudin-1 and -2: novel integral membrane proteins localizing at tight junctions with no sequence similarity to occludin. J Cell Biol 1998;141(7):1539-1550.

55. Furuse M, Tsukita S. Claudins in occluding junctions of humans and flies. Trends Cell Biol 2006;16(4):181-188.

56. Tsukita S, Furuse M. Occludin and claudins in tight-junction strands: leading or supporting players? Trends Cell Biol 1999;9(7):268-273.

57. Tsukita S, Furuse M. Overcoming barriers in the study of tight junction functions: from occludin to claudin. Genes Cells 1998;3(9):569-573.

58. Van Itallie CM, Anderson JM. Claudins and epithelial paracellular transport. Annu Rev Physiol 2006;68:403-429.

59. Tsukita S, Furuse M. [Identification of two distinct types of four-transmembrane domain proteins, occludin and claudins: towards new physiology in paracellular pathway]. Seikagaku 2000;72(3):155-162.

60. Tsukita S, Furuse M. The structure and function of claudins, cell adhesion molecules at tight junctions. Ann NY Acad Sci 2000;915:129-135.

61. Saitou M, Fujimoto K, Doi Y et al. Occludin-deficient embryonic stem cells can differentiate into polarized epithelial cells bearing tight junctions. J Cell Biol 1998;141(2):397-408.

62. Furuse M, Hirase T, Itoh M et al. Occludin: a novel integral membrane protein localizing at tight junctions. J Cell Biol 1993;123(6 Pt 2):1777-1788.

63. Furuse M, Itoh M, Hirase T et al. Direct association of occludin with ZO-1 and its possible involvement in the localization of occludin at tight junctions. J Cell Biol 1994;127(6 Pt 1):1617-1626.

64. McCarthy KM, Skare IB, Stankewich MC et al. Occludin is a functional component of the tight junction. J Cell Sci 1996;109(Pt 9):2287-2298.

65. Saitou M, Furuse M, Sasaki H et al. Complex phenotype of mice lacking occludin, a component of tight junction strands. Mol Biol Cell 2000;11(12):4131-4142.

66. Schulzke JD, Gitter AH, Mankertz J et al. Epithelial transport and barrier function in occludin-deficient mice. Biochim Biophys Acta 2005;1669(1):34-42.

67. Nusrat A, Chen JA, Foley CS et al. The coiled-coil domain of occludin can act to organize structural and functional elements of the epithelial tight junction. J Biol Chem 2000;275(38):29816-29822.

68. Medina R, Rahner C, Mitic LL, Anderson JM, Van Itallie CM. Occludin localization at the tight junction requires the second extracellular loop. J Membr Biol 2000;178(3):235-247.

69. Matter K, Balda MS. Biogenesis of tight junctions: the C-terminal domain of occludin mediates basolateral targeting. J Cell Sci 1998;111(Pt 4):511-519.

70. Matter K, Balda MS. Occludin and the functions of tight junctions. Int Rev Cytol 1999;186:117-146.

71. Ikenouchi J, Furuse M, Furuse K, Sasaki H, Tsukita S, Tsukita S. Tricellulin constitutes a novel barrier at tricellular contacts of epithelial cells. J Cell Biol 2005;171(6):939-945.

72. Van Itallie CM, Anderson JM. The molecular physiology of tight junction pores. Physiology (Bethesda) 2004;19:331-338.

73. Madara JL, Parkos C, Colgan S, Nusrat A, Atisook K, Kaoutzani P. The movement of solutes and cells across tight junctions. Ann NY Acad Sci 1992;664:47-60.

74. Madara JL. Regulation of the movement of solutes across tight junctions. Annu Rev Physiol 1998;60:143-159.

75. Cereijido M, Valdes J, Shoshani L, Contreras RG. Role of tight junctions in establishing and maintaining cell polarity. Annu Rev Physiol 1998;60:161-177.

76. Cereijido M, Shoshani L, Contreras RG. Molecular physiology and pathophysiology of tight junctions. I. Biogenesis of tight junctions and epithelial polarity. Am J Physiol Gastrointest Liver Physiol 2000;279(3):G477-G482.

77. Balda MS, Matter K. Tight junctions and the regulation of gene expression. Biochim Biophys Acta 2009;1788(4):761-767.

78. Matter K, Aijaz S, Tsapara A, Balda MS. Mammalian tight junctions in the regulation of epithelial differentiation and proliferation. Curr Opin Cell Biol 2005;17(5):453-458.

79. Baumgart DC, Dignass AU. Intestinal barrier function. Curr Opin Clin Nutr Metab Care 2002;5(6):685-694.

80. Marchiando AM, Graham WV, Turner JR. Epithelial barriers in homeostasis and disease. Annu Rev Pathol 2010;5:119-144.

81. Cereijido M, Contreras RG, Flores-Benitez D et al. New diseases derived or associated with the tight junction. Arch Med Res 2007;38(5):465-478.

82. Edelblum KL, Turner JR. The tight junction in inflammatory disease: communication breakdown. Curr Opin Pharmacol 2009;9(6):715-720.

83. Hollander D. Intestinal permeability, leaky gut, and intestinal disorders. Curr Gastroenterol Rep 1999;1(5):410-416.

84.   Su L, Shen L, Clayburgh DR et al. Targeted epithelial tight junction dysfunction causes immune activation and contributes to development of experimental colitis. Gastroenterology 2009;136(2):551-563.

85.   Anderson JM, Balda MS, Fanning AS. The structure and regulation of tight junctions. Curr Opin Cell Biol 1993;5(5):772-778.

86.   Johnson LN. The regulation of protein phosphorylation. Biochem Soc Trans 2009;37(Pt 4):627-641.

87.   Matter K, Balda MS. Signalling to and from tight junctions. Nat Rev Mol Cell Biol 2003;4(3):225-236.

88.   Balda MS, Gonzalez-Mariscal L, Contreras RG et al. Assembly and sealing of tight junctions: possible participation of G-proteins, phospholipase C, protein kinase C and calmodulin. J Membr Biol 1991;122(3):193-202.

89.   Balda MS, Gonzalez-Mariscal L, Matter K, Cereijido M, Anderson JM. Assembly of the tight junction: the role of diacylglycerol. J Cell Biol 1993;123(2):293-302.

90.   Kohler K, Louvard D, Zahraoui A. Rab13 regulates PKA signaling during tight junction assembly. J Cell Biol 2004;165(2):175-180.

91.   Denker BM, Saha C, Khawaja S, Nigam SK. Involvement of a heterotrimeric G protein alpha subunit in tight junction biogenesis. J Biol Chem 1996;271(42):25750-25753.

92.   Dodane V, Kachar B. Identification of isoforms of G proteins and PKC that colocalize with tight junctions. J Membr Biol 1996;149(3):199-209.

93.   Shen L, Black ED, Witkowski ED et al. Myosin light chain phosphorylation regulates barrier function by remodeling tight junction structure. J Cell Sci 2006;119(Pt 10):2095-2106.

94.   Takaishi K, Sasaki T, Kotani H, Nishioka H, Takai Y. Regulation of cell-cell adhesion by rac and rho small G proteins in MDCK cells. J Cell Biol 1997;139(4):1047-1059.

95.   Hasegawa H, Fujita H, Katoh H et al. Opposite regulation of transepithelial electrical resistance and paracellular permeability by Rho in Madin-Darby canine kidney cells. J Biol Chem 1999;274(30):20982-20988.

96.   Li D, Mrsny RJ. Oncogenic Raf-1 disrupts epithelial tight junctions via downregulation of occludin. J Cell Biol 2000;148(4):791-800.

97. Chen Y, Lu Q, Schneeberger EE, Goodenough DA. Restoration of tight junction structure and barrier function by down-regulation of the mitogen-activated protein kinase pathway in ras-transformed Madin-Darby canine kidney cells. Mol Biol Cell 2000;11(3):849-862.

98. Boettner B, Govek EE, Cross J, Van AL. The junctional multidomain protein AF-6 is a binding partner of the Rap1A GTPase and associates with the actin cytoskeletal regulator profilin. Proc Natl Acad Sci U S A 2000;97(16):9064-9069.

99. Fujita H, Katoh H, Hasegawa H et al. Molecular decipherment of Rho effector pathways regulating tight-junction permeability. Biochem J 2000;346 Pt 3:617-622.

100. Hirase T, Kawashima S, Wong EY et al. Regulation of tight junction permeability and occludin phosphorylation by Rhoa-p160ROCK-dependent and -independent mechanisms. J Biol Chem 2001;276(13):10423-10431.

101. Tepass U, Theres C, Knust E. crumbs encodes an EGF-like protein expressed on apical membranes of Drosophila epithelial cells and required for organization of epithelia. Cell 1990;61(5):787-799.

102. Lechward K, Awotunde OS, Swiatek W, Muszynska G. Protein phosphatase 2A: variety of forms and diversity of functions. Acta Biochim Pol 2001;48(4):921-933.

103. Nunbhakdi-Craig V, Machleidt T, Ogris E, Bellotto D, White CL, III, Sontag E. Protein phosphatase 2A associates with and regulates atypical PKC and the epithelial tight junction complex. J Cell Biol 2002;158(5):967-978.

104. Seth A, Sheth P, Elias BC, Rao R. Protein phosphatases 2A and 1 interact with occludin and negatively regulate the assembly of tight junctions in the CACO-2 cell monolayer. J Biol Chem 2007;282(15):11487-11498.

105. Sheth P, Samak G, Shull JA, Seth A, Rao R. Protein phosphatase 2A plays a role in hydrogen peroxide-induced disruption of tight junctions in Caco-2 cell monolayers. Biochem J 2009;421(1):59-70.

106. Clarke H, Soler AP, Mullin JM. Protein kinase C activation leads to dephosphorylation of occludin and tight junction permeability increase in LLC-PK1 epithelial cell sheets. J Cell Sci 2000;113(Pt 18):3187-3196.

107. Suzuki T, Elias BC, Seth A et al. PKC eta regulates occludin phosphorylation and epithelial tight junction integrity. Proc Natl Acad Sci U S A 2009;106(1):61-66.

108. Atkinson KJ, Rao RK. Role of protein tyrosine phosphorylation in acetaldehyde-induced disruption of epithelial tight junctions. Am J Physiol Gastrointest Liver Physiol 2001;280(6):G1280-G1288.

109. Chen YH, Lu Q, Goodenough DA, Jeansonne B. Nonreceptor tyrosine kinase c-Yes interacts with occludin during tight junction formation in canine kidney epithelial cells. Mol Biol Cell 2002;13(4):1227-1237.

110. Basuroy S, Sheth P, Kuppuswamy D, Balasubramanian S, Ray RM, Rao RK. Expression of kinase-inactive c-Src delays oxidative stress-induced disassembly and accelerates calcium-mediated reassembly of tight junctions in the Caco-2 cell monolayer. J Biol Chem 2003;278(14):11916-11924.

111. Sakakibara A, Furuse M, Saitou M, Ando-Akatsuka Y, Tsukita S. Possible involvement of phosphorylation of occludin in tight junction formation. J Cell Biol 1997;137(6):1393-1401.

112. Rao RK, Basuroy S, Rao VU, Karnaky Jr KJ, Gupta A. Tyrosine phosphorylation and dissociation of occludin-ZO-1 and E-cadherin-beta-catenin complexes from the cytoskeleton by oxidative stress. Biochem J 2002;368(Pt 2):471-481.

113. Elias BC, Suzuki T, Seth A et al. Phosphorylation of Tyr-398 and Tyr-402 in occludin prevents its interaction with ZO-1 and destabilizes its assembly at the tight junctions. J Biol Chem 2009;284(3):1559-1569.

114. Basuroy S, Seth A, Elias B, Naren AP, Rao R. MAPK interacts with occludin and mediates EGF-induced prevention of tight junction disruption by hydrogen peroxide. Biochem J 2006;393(Pt 1):69-77.

115. Basuroy S, Dunagan M, Sheth P, Seth A, Rao RK. Hydrogen peroxide activates focal adhesion kinase and c-Src by a phosphatidylinositol 3 kinase-dependent mechanism and promotes cell migration in Caco-2 cell monolayers. Am J Physiol Gastrointest Liver Physiol 2010;299(1):G186-G195.

116. Izumi Y, Hirose T, Tamai Y et al. An atypical PKC directly associates and colocalizes at the epithelial tight junction with ASIP, a mammalian homologue of Caenorhabditis elegans polarity protein PAR-3. J Cell Biol 1998;143(1):95-106.

117. Mullin JM, Kampherstein JA, Laughlin KV et al. Overexpression of protein kinase C-delta increases tight junction permeability in LLC-PK1 epithelia. Am J Physiol 1998;275(2 Pt 1):C544-C554.

118. Chen ML, Pothoulakis C, LaMont JT. Protein kinase C signaling regulates ZO-1 translocation and increased paracellular flux of T84 colonocytes exposed to Clostridium difficile toxin A. J Biol Chem 2002;277(6):4247-4254.

119. Sheth P, Basuroy S, Li C, Naren AP, Rao RK. Role of phosphatidylinositol 3-kinase in oxidative stress-induced disruption of tight junctions. J Biol Chem 2003;278(49):49239-49245.

120. Aggarwal S, Suzuki T, Taylor WL, Bhargava A, Rao R. Contrasting Effects of ERK on Tight Junction Integrity in Differentiated and Under-Differentiated Caco-2 Cell Monolayers. Biochem J 2010.

121. Wong V. Phosphorylation of occludin correlates with occludin localization and function at the tight junction. Am J Physiol 1997;273(6 Pt 1):C1859-C1867.

122. Stevenson BR, Goodenough DA. Zonulae occludentes in junctional complex-enriched fractions from mouse liver: preliminary morphological and biochemical characterization. J Cell Biol 1984;98(4):1209-1221.

123. Cordenonsi M, Turco F, D'Atri F et al. Xenopus laevis occludin. Identification of in vitro phosphorylation sites by protein kinase CK2 and association with cingulin. Eur J Biochem 1999;264(2):374-384.

124. Hirai T, Chida K. Protein kinase Czeta (PKCzeta): activation mechanisms and cellular functions. J Biochem 2003;133(1):1-7.

125. Andreeva AY, Krause E, Muller EC, Blasig IE, Utepbergenov DI. Protein kinase C regulates the phosphorylation and cellular localization of occludin. J Biol Chem 2001;276(42):38480-38486.

126. Helfrich I, Schmitz A, Zigrino P et al. Role of aPKC isoforms and their binding partners Par3 and Par6 in epidermal barrier formation. J Invest Dermatol 2007;127(4):782-791.

127. Hidalgo IJ, Raub TJ, Borchardt RT. Characterization of the human colon carcinoma cell line (Caco-2) as a model system for intestinal epithelial permeability. Gastroenterology 1989;96(3):736-749.

128. Zucco F, Batto AF, Bises G et al. An inter-laboratory study to evaluate the effects of medium composition on the differentiation and barrier function of Caco-2 cell lines. Altern Lab Anim 2005;33(6):603-618.

129. Sambuy Y, de Angelis I, Ranaldi G, Scarino ML, Stammati A, Zucco F. The Caco-2 cell line as a model of the intestinal barrier: influence of cell and culture-related factors on Caco-2 cell functional characteristics. Cell Biol Toxicol 2005;21(1):1-26.

130. Artursson P, Karlsson J. Correlation between oral drug absorption in humans and apparent drug permeability coefficients in human intestinal epithelial (Caco-2) cells. Biochem Biophys Res Commun 1991;175(3):880-885.

131. Irvine JD, Takahashi L, Lockhart K et al. MDCK (Madin-Darby canine kidney) cells: A tool for membrane permeability screening. J Pharm Sci 1999;88(1):28-33.

132. Matter K, Balda MS. Functional analysis of tight junctions. Methods 2003;30(3):228-234.

133. Middleton E. The molecular configuration of inulin: implications for ultrafiltration theory and glomerular permeability. J Membr Biol 1977;34(1):93-101.

134. Korzeniewski C, Callewaert DM. An enzyme-release assay for natural cytotoxicity. J Immunol Meth 1983;64(3):313-320.

135. Decker T, Lohmann-Matthes ML. A quick and simple method for the quantitation of lactate dehydrogenase release in measurements of cellular cytotoxicity and tumor necrosis factor (TNF) activity. J Immunol Meth 1988;115(1):61-69.

136. Berridge MV, Herst PM, Tan AS. Tetrazolium dyes as tools in cell biology: new insights into their cellular reduction. Biotechnol Annu Rev 2005;11:127-152.

137. Nigam SK, Rodriguez-Boulan E, Silver RB. Changes in intracellular calcium during the development of epithelial polarity and junctions. Proc Natl Acad Sci U S A 1992;89(13):6162-6166.

138. Stuart RO, Sun A, Panichas M, Hebert SC, Brenner BM, Nigam SK. Critical role for intracellular calcium in tight junction biogenesis. J Cell Physiol 1994;159(3):423-433.

139. Harhaj NS, Antonetti DA. Regulation of tight junctions and loss of barrier function in pathophysiology. Int J Biochem Cell Biol 2004;36(7):1206-1237.

140. Madara JL, Nash S, Moore R, Atisook K. Structure and function of the intestinal epithelial barrier in health and disease. Monogr Pathol 1990;(31):306-324.

141. Akimoto K, Mizuno K, Osada S et al. A new member of the third class in the protein kinase C family, PKC lambda, expressed dominantly in an undifferentiated mouse embryonal carcinoma cell line and also in many tissues and cells. J Biol Chem 1994;269(17):12677-12683.

142. Capaldo CT, Macara IG. Depletion of E-cadherin disrupts establishment but not maintenance of cell junctions in Madin-Darby canine kidney epithelial cells. Mol Biol Cell 2007;18(1):189-200.

143. Farshori P, Kachar B. Redistribution and phosphorylation of occludin during opening and resealing of tight junctions in cultured epithelial cells. J Membr Biol 1999;170(2):147-156.

144. Koizumi J, Kojima T, Ogasawara N et al. Protein kinase C enhances tight junction barrier function of human nasal epithelial cells in primary culture by transcriptional regulation. Mol Pharmacol 2008;74(2):432-442.

145. Chou MM, Hou W, Johnson J et al. Regulation of protein kinase C zeta by PI 3-kinase and PDK-1. Curr Biol 1998;8(19):1069-1077.

146. Sheth P, Seth A, Thangavel M, Basuroy S, Rao RK. Epidermal growth factor prevents acetaldehyde-induced paracellular permeability in Caco-2 cell monolayer. Alco Clin Exp Res 2004;28(5):797-804.

147. Mitic LL, Anderson JM. Molecular architecture of tight junctions. Ann Rev Phys 1998; 60:121-142.

148   Niessen CM. Tight Junctions/Adherens Junctions: Basic Structure and Function. Retrieved from http://www.nature.com/jid/journal/v127/n11/full/5700865a.html. Accessed on 06 May 2011.

149. Schneeberger E.E., Lynch R.D. The tight junction: a multifunctional complex. Am J Physiol 2004;286:1213-1228.

150. Barrier Forming Tissue.Retrieved from http://www.nanoanalytics.de/en/hardwareproducts/cellzscope/howitworks/chapter01/index.php. Accessed on 06 May 2011.

151. Chida K, Hirai T. Protein kinase C zeta (PKCζ): activation mechanisms and cellular functions. J Biochem (Tokyo) 2003 Mar;133(3):395.

152. *Millicell®–ERS User Guide* P17304, Rev. C, 2 (2007). Millipore Corporation Billerica, MA.

153. *Millicell®–ERS User Guide* P17304, Rev. C, 7 (2007). Millipore Corporation Billerica, MA.

154. The LDH Assay. Retrieved from http://www.gbiosciences.com/CytoscanLDHCytotoxicityAssayKit.aspx. Accessed on 06 May 2011.

155. pRNATin-H1.2/Neo vector. Retrieved from http://www.genscript.com/product_001/marker/code/SD1223/siRNA%20Expression%20Vector/pRNATin_H1_2_Neo/SD1223.html. Accessed on 06 May 2011.

VITA

Suneet Kumar Jain was born in Delhi, India in the year 1978. He attended high school in Ranikhet, India and graduated from high school in 1995. He joined B.J. Medical College, University of Pune in and obtained his Bachelor in Medicine, Bachelor in Surgery (M.B.B.S.) degree in 2003. In August 2004, he joined the Integrated Program in Biomedical Sciences (IPBS) at the University of Tennessee Health Sciences Center (UTHSC) Memphis to work for his Ph.D. A member of American Gastroenterology Association (AGA), he has attended 2007 Digestive Disease Week where his poster was selected as 'poster of distinction.'